Measuring Indoor Air Quality

PRINCIPLES AND TECHNIQUES IN THE ENVIRONMENTAL SCIENCES

Decaying Lakes: The Origins and Control of Cultural Eutrophication
B. Henderson-Sellers and H. R. Markland

Measuring Indoor Air Quality: A Practical Guide
John E. Yocom and Sharon M. McCarthy

Measuring Indoor Air Quality
A Practical Guide

John E. Yocom, P.E., C.I.H.
Environmental Consultant
West Simsbury, Connecticut, U.S.A.

and

Sharon M. McCarthy, Ph.D.
Sigma Research Corporation
Westford, Massachusetts, U.S.A.

JOHN WILEY & SONS
Chichester · New York · Brisbane · Toronto · Singapore

Copyright © 1991 by John Wiley & Sons Ltd.,
Baffins Lane, Chichester,
West Sussex PO19 1UD, England

Other Wiley Editorial Offices

John Wiley & Sons, Inc., 605 Third Avenue,
New York, NY 10158-0012, USA

Jacaranda Wiley Ltd, G.P.O. Box 859, Brisbane,
Queensland 4001, Australia

John Wiley & Sons (Canada) Ltd, 5353 Dundas Road West, Fourth Floor,
Etobicoke, Ontario M9B 6H8, Canada

John Wiley & Sons (SEA) Pte Ltd, 37 Jalan Pemimpin 05-04,
Block B, Union Industrial Building, Singapore 2057

Library of Congress Cataloging-in-Publication Data:

Yocom, J. E. (John E.)
Measuring indoor air quality: a practical guide / by John E. Yocom and
Sharon M. McCarthy.
p. cm. — (Principles and techniques in the environmental
sciences)
Includes bibliographical references and index.
ISBN 0-471-90728-6 (ppc)
1. Indoor air pollution—Measurement. 2. Air quality—
Measurement. I. McCarthy, Sharon M. II. Title. III. Series.
TD890.Y63 1991
628.5′35—dc20
91–9385
CIP

British Library Cataloguing in Publication Data:

Yocom, John E.
Measuring indoor air quality: a practical guide.
– (Principles and techniques in the
environmental sciences)
I. Title II. McCarthy, Sharon M. III. Series
697.9028

ISBN 0-471-90728-6

Typeset in 10/12pt Times
Printed in Great Britain by Courier International Ltd, East Kilbride

Contents

Series Preface

Most environmental texts concentrate on the 'what' but not on the 'how' and on theories rather than practice. In this series on Principles and Techniques in the Environmental Sciences, therefore, we encouraged authors to focus on explaining how their analytical framework could be utilized in solving real-world problems.

Our audience is intended to include college students and practising professionals eager to update their expertise in a specialized practical area. Within the series, some books will be in the form of texts adoptable for class use while others will function more readily as professional handbooks. In all cases, the emphasis will be on the *application* of knowledge, although naturally a concise description of the relevant disciplines will be a critical component of the discourse.

We trust that such an application-oriented series of books will be both welcome and useful to environmental professionals.

B. Henderson-Sellers
University of New South Wales

W.Y.B. Chang
University of Michigan

March 1991

To
Betty and Jim

Preface

It has been well recognized for 20 or 25 years that people, at least in the developed countries, spend most of their lives indoors where air quality is quite different and often much worse than that outdoors. Yet the emphasis on air quality improvement is still on controlling sources of outdoor air pollution. Nevertheless, there has been a rapid expansion of interest over the past few years in indoor air quality and its contribution to total human exposure to air pollutants.

This book has had a long gestation period (1982–90) because the authors, working as environmental consultants, had difficulty in finding uninterrupted periods for preparation of the manuscript between work, travel, job changes, marriage, and fragmented home life, all in the face of powerful urges to procrastinate. One might rationalize that this worked to our advantage since the period over which the book was prepared saw an accelerating interest in indoor air quality which has produced a large number of indoor air quality studies and many new developments in monitoring techniques. Thus, a book such as this can only hope to inform the reader of past and present developments at one point in time, and present the authors' view of the relative importance of various monitoring techniques and their opinions on where the field is or should be heading. We hope that readers will gain an improved understanding of indoor air quality and how to monitor it, and that they will gain from the experiences of others and will not attempt to repeat prior studies.

Although the time we spent writing was our own, we are indebted to the companies for whom we work or have worked (TRC Environmental Consultants, Inc., E. C. Jordan Co., and Sigma Research Corporation) for typing, editing, copying, and limited secretarial services. We are also grateful to our spouses for their patience and encouragement. Special thanks are given to Ms. Sandy Kaleda of TRC for performing the bulk of the typing.

John E. Yocom
West Simsbury, Connecticut

Sharon M. McCarthy
Westford, Massachusetts

November 1990

CHAPTER 1

Introduction

The quality of air to which an individual is exposed may be defined in two ways: qualitatively by descriptors such as odor perceived by the individual or quantitatively by the relative concentrations of the various components of the air as determined by chemical or physical measurements. When one enters a building he or she will immediately perceive any significant differences in temperature and relative humidity as compared with conditions outdoors, but beyond these physical characteristics, the indoor space will usually have an entirely different odor, leading the individual to the conclusion that indoor air quality may be quite different chemically from outdoor air quality. This is not to say that the concentrations of the principal constituents of the atmosphere are significantly different indoors than outdoors. Table 1.1 presents a typical analysis of the earth's atmosphere. Except under highly unusual circumstances, such as those occurring in a tightly built home in which a number of unvented combustion devices are operated, the oxygen concentration indoors will approximate that shown in Table 1.1 for outdoor air. In fact, oxygen concentrations in normally occupied indoor spaces will seldom be depressed as much as a percent even under the most extreme conditions. While data are limited, deaths in homes with unvented or improperly vented combustion

Table 1.1 Composition of dry, unpolluted outdoor air at ground level

Constituent	Fraction volume abundance	
	Percent	*Part per million by volume (ppm)*
Nitrogen	78.08	
Oxygen	20.95	
Argon	0.93	
Carbon dioxide	0.03	
Neon		18.2
Helium		5.2
Methane		1.5
Krypton		1.1
Nitrous oxide		0.5
Carbon monoxide (variable)		0.1 to 0.2
Xenon		0.1

systems are most often the result of elevated carbon monoxide concentrations and not oxygen depletion.

The carbon dioxide (CO_2) concentration inside of occupied buildings will invariably be higher than typical values outdoors as a result of human respiration. Indoor concentrations of CO_2 five times those found outdoors are not uncommon in poorly ventilated, occupied spaces. Therefore, the concerns about indoor air quality are related not so much to changes in the gross constituents but to the relatively small additions of a wide variety of compounds from indoor sources and activities and from pollutants in the outdoor ambient atmosphere.

1.1 Outdoor Air Monitoring

Since 1970 a considerable amount of air monitoring has been carried out to measure concentrations of air pollutants in the outdoor ambient atmosphere. In the United States this monitoring has been necessary to determine the status of compliance with National Ambient Air Quality Standards (NAAQS). These standards were developed by the U.S. Environmental Protection Agency (U.S. EPA) under the terms of the Clean Air Act (CAA). The primary purpose of this Act and its subsequent amendments is to improve ambient air quality by reducing emissions of a group of pollutants capable of impairing human health if concentrations of the pollutants are sufficiently high. The pollutants are: sulfur dioxide (SO_2), total suspended particulate matter (TSP), inhalable particulate matter (PM_{10}) carbon monoxide (CO), nitrogen dioxide (NO_2), photochemical oxidants or ozone (O_3), and lead (Pb). Table 1.2 presents the current U.S. ambient outdoor air quality standards for these pollutants. Note that for most pollutants there are "primary" standards which are based on human health effects and secondary standards which are based on welfare effects such as vegetation damage or soiling. These pollutants are often called "criteria pollutants" since the CAA required that criteria or justification for the control must be published before air quality standards are set. The selected pollutants are released from a variety of outdoor sources. The emphasis on outdoor air pollution and its measurement has been built into the EPA's regulations implementing the CAA. These regulations define ambient air as "that portion of the atmosphere, external to buildings, to which the general public has access" (U.S. Code of Federal Regulations, Part 50, 1989).

Tracking the results of the many programs in the United States to monitor concentrations of the "criteria pollutants" over the past 20 years has shown conclusively that emission controls have been quite effective in improving outdoor air quality. However, epidemiological studies have not been able to show that this improved outdoor air quality has produced a similar improvement in the health of the general public. One reason is that people spend most of their time indoors where air quality is quite different than it is outdoors. Pollutants indoors may differ in type and concentration from outdoor pollutants. The concentrations of some pollutants

Table 1.2 The U.S. national ambient air quality standards

Pollutant	Averaging time	Primary standards	Secondary standards
Sulfur Dioxide (SO₂)	Annual arithmetic mean	$80 \ \mu g/m^3$ (0.03 ppm)	
	24-hour*	$365 \ \mu g/m^3$ (0.140 ppm)	
	3-hour*		$1300 \ \mu g/m^3$ (0.50 ppm)
Total suspended	Annual geometric mean	$75 \ \mu g/m^3$ $\quad 60 \ \mu g/m^3$	$150 \ \mu g/m^3$
Particulate matter	24-hour*	$260 \ \mu g/m^3$	
(TSP)†			
Inhalable particulate	Annual arithmetic mean	$50 \ \mu g/m^3$	$50 \ \mu g/m^3$
Matter (PM₁₀)	24-hour	$150 \ \mu g/m^3$	$150 \ \mu g/m^3$
Carbon monoxide (CO)	8 hour*	$10 \ \mu g/m^3$ (9 ppm)	
	1 hour*	$40 \ \mu g/m^3$ (35 ppm)	
Ozone (O₃)	1 hour‡	$235 \ \mu g/m^3$ (0.12 ppm)	$235 \ \mu g/m^3$ (0.12 ppm)
Nitrogen dioxide (NO₂)	Annual arithmetic mean	$100 \ \mu g/m^3$ (0.053 ppm)	$100 \ \mu g/m^3$ (0.053 ppm)
Lead (Pb)	Calendar quarter average	$1.5 \ \mu g/m^3$	$1.5 \ \mu g/m^3$

* Maximum concentration not to be exceeded more than once per year.

† This standard is being phased out as EPA implements the inhalable (PM₁₀) standard.

‡ The standard is attained when the expected number of days per calendar year with maximum hourly average concentrations above 0.12 ppm is equal to or less than one.

with predominantly outdoor sources (e.g., SO_2) are reduced as they penetrate the indoor environment. On the other hand, concentrations of pollutants with both outdoor and indoor sources (e.g., NO_2) may be higher indoors than outdoors when indoor sources are present. Concentrations of pollutants with predominantly indoor sources (e.g., formaldehyde) are invariably higher indoors than outdoors.

1.2 Indoor Air Quality Monitoring

In this book the term "indoor air" will apply to all indoor environments except indoor industrial (e.g., factory) occupational environments. Such industrial settings are covered by occupational exposure standards used by industrial hygienists. Thus, the scope of this book includes the indoor air in residences, office buildings, and public buildings (e.g., schools, hospitals, theaters, restaurants, etc.). Of course, office buildings represent occupational settings in which there is an employer–employee relationship implying that occupational indoor air quality standards apply. However, the concentrations of air contaminants in such locations tend to be in the same range as those outdoors which are much lower than in industrial locations where occupational indoor air quality standards commonly apply. Nevertheless, workers in office buildings complain about indoor air quality when concentrations of pollutants are orders of magnitude below occupational standards. Thus, there is a need to investigate such situations. The approach to such investigations will be discussed in Chapter 2.

The recognition that indoor exposure to air pollutants comprises an important component of total exposure to air pollution has developed only since about 1965 with most of the research having been carried out since about 1975. The first indoor air quality studies dealt with the relationships between indoor and outdoor pollutant concentrations since the primary concern was determining the degree of penetration of outdoor pollutants indoors. The earliest significant study of indoor–outdoor air quality levels was carried out in the Netherlands by Biersteker *et al*. (1965). In this study SO_2 was measured both indoors and outdoors at 60 Rotterdam homes; indoor concentrations of SO_2, a reactive pollutant of outdoor generation, were always lower than those outdoors. In the first major U.S. indoor-outdoor air quality study, Yocom *et al*. (1971) confirmed this behavior of SO_2 but found that gas stoves can produce elevated levels of CO indoors as compared with outdoors. Subsequent studies which were summarized and critically reviewed by Yocom (1982) included additional pollutants and increased the emphasis on pollutants generated primarily indoors.

Because of the differences between indoor and outdoor air quality and the predominance of time spent indoors (over 90 percent for most people in developed countries), it is imperative that data be gathered on indoor exposures to pollutants if there is serious concern about the health effects of air pollutants. The emphasis that the U.S. EPA has historically placed upon outdoor air quality is based on the traditional belief that pollutants generated by large outdoor sources (power plants, industries, automobiles, etc.) are of primary concern with regard to human health. Table 1.3, from a National Academy of Sciences report (NAS, 1981), shows the types of air pollutants of importance indoors and categorizes the pollutants in terms of their origins. Figure 1.1 is a diagram of an office building in a metropolitan area

Table 1.3 Typical sources of some pollutants grouped by origin (NAS, 1981)

Pollutants	Sources
Group I—Sources predominantly outdoor:	
Sulfur oxides (gases, particles)	Fuel combustion, smelters
Ozone	Photochemical reactions
Pollens	Trees, grass, weeds, plants
Lead, manganese	Automobiles
Calcium, chlorine, silicon, cadmium	Suspension of soils or industrial emission
Organic substances	Petrochemical solvents, natural sources, vaporization of unburned fuels
Group II—Sources both indoor and outdoor:	
Nitric oxide, nitrogen dioxide	Fuel-burning
Carbon monoxide	Fuel-burning
Carbon dioxide	Metabolic activity, combustion
Particles	Resuspension, condensation of vapors and combustion products
Water vapor	Biologic activity, combustion, evaporation
Organic substances	Volatilization, combustion, paint, metabolic action, pesticides, insecticides, fungicides
Spores	Fungi, molds
Group III—Sources predominantly indoor:	
Radon	Underlying rock and soil, building construction materials (concrete, stone), water
Formaldehyde	Particle board, insulation, furnishings, tobacco smoke
Asbestos, mineral, and synthetic fibers	Fire-retardant, acoustic, thermal, or electric insulation
Organic substances	Adhesives, solvents, cooking, cosmetics
Ammonia	Metabolic activity, cleaning products
Polycyclic hydrocarbons, arsenic, nicotine, acrolein, etc.	Tobacco smoke
Mercury	Fungicides in paints, spills in dental care facilities or laboratories, thermometer breakage
Aerosols	Consumer products
Viable organisms	Infections
Allergens	House dust, animal dander

6

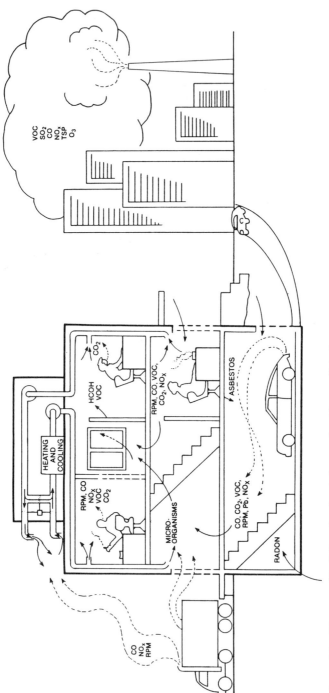

Figure 1.1 Diagram of building showing indoor and outdoor pollutant sources

showing indoor and outdoor sources of some of these pollutants. Particulate matter is ubiquitous both indoors and outdoors, but the particles generated indoors may be quite different chemically and physically than those generated indoors. Furthermore, outdoor ambient particulate matter contains a significant fraction of large diameter particles which do not readily penetrate indoors. Certain inorganic gases, e.g., SO_2 and O_3, produced by outdoor sources, are reactive and are effectively reduced indoors due to physical and chemical reactions with surfaces. Organic gases and vapors are also present in both the outdoor and indoor atmospheres. Some are produced only outdoors, some are produced only indoors, and others are of both indoor and outdoor origin. From these examples alone, it should be obvious that the accurate monitoring of indoor air quality will help to determine the concentrations of pollutants that characterize human exposure. The results of indoor air quality monitoring studies coupled with measurement of emissions from sources both indoors and outdoors will provide the basis for appropriate action to control many problems of excess exposure to pollutants.

1.3 Factors that Influence Indoor Air Quality

Many factors affect indoor air quality and all of them may need to be dealt with in a monitoring program. These factors have been discussed by Yocom (1982) and include the following:

- Outdoor air quality.
- Indoor generation of pollutants.
- Pollution depletion mechanisms (deposition and chemical decay).
- Meteorological factors (affects air exchange).
- Permeability of structure (affects air exchange).
- Ventilation measures (affects air exchange).

The following simple mass balance model shows how the indoor concentration of a given pollutant is affected quantitatively by the principal influencing factors:

$$\frac{dC_i}{dt} = aC_o - aC_i - DC_i + S \tag{1.1}$$

where: C_i is indoor concentration, C_o is outdoor concentration, a is air exchange rate, D is pollutant decay rate, and S is indoor source strength per unit volume.

The steady state condition is shown by setting the left hand side of the equation equal to zero producing the following relationship:

$$C_i = \frac{aC_o + S}{a + D} \tag{1.2}$$

These equations will be useful in the chapters that follow describing techniques for monitoring indoor concentrations and the factors (independent variables) that influence indoor air quality.

In planning an indoor air monitoring program, one must also consider the purposes for which the indoor air quality data are required. The types of studies requiring indoor air quality measurements include, but are certainly not limited to, the following:

- Characterizing indoor–outdoor relationships for several types of structures and pollutants.
- Measuring the effect of indoor sources (e.g., unvented gas stoves) on indoor air quality.
- Determining the effect of weatherization of structures on indoor air quality.
- Measuring pollutant distribution within structures.
- Measuring pollutant emission rates of indoor sources.
- Determining human exposure to pollutants and relating the monitoring data to health effects.
- Determining the cause of complaints about indoor air quality from building occupants (Sick Building Syndrome).

The objectives of a study will affect profoundly the monitoring approaches and sampling approaches required. For example, a study of the effect of indoor sources such as unvented combustion devices on indoor air quality may require rapid-response, continuous analyzers for combustion products such as CO or NO_2. The duration of the program is dictated by the length of time required to obtain a representative sampling of source use and seasonal meteorological conditions and other factors that affect the results. On the other hand, monitoring total human exposure as part of an epidemiological study requires not only indoor monitoring of both long-term average and short-term peak concentrations, but also personal monitoring that depicts total exposure to the individuals carrying or wearing the monitors. An indoor air quality and total exposure study as part of a prospective epidemiological program must be carried out over an extended period. For example, in a longitudinal program, such as the Harvard Six Cities Study (Dockery and Spengler, 1981), monitoring must proceed for many years.

1.3.1 Types of Monitoring Systems

Most indoor air quality monitoring systems have been derived from equipment used to sample occupational exposures in industrial plants or outdoor ambient atmospheres. Many of the methods used for industrial occupational settings (industrial hygiene methods) are not sufficiently sensitive for the indoor settings of concern such as office buildings and homes. These industrial hygiene sampling methods were designed to measure the relatively high concentrations of pollutants that could be permitted in exposure of robust factory workers 8 hours per day, 5 days per week. Indoor air quality methods must be sensitive enough to measure accurately the low concentrations of pollutants that may affect extremely sensitive populations (e.g., infants, the aged, and the infirm) exposed 24 hours a day and 7 days per week.

Methods developed for outdoor ambient monitoring operate generally in the concentration ranges of most indoor air quality situations, but the number of pollutants for which such methods are available is limited, and often there are difficulties in using some of this instrumentation indoors. For example, the high volume air sampler (Hi-Vol) which had earlier been the standard method of measuring total suspended particulate matter (TSP) in the U.S. is too noisy to be operated in an occupied space, and the large volume of air which it filters to collect particulate matter is likely to change the indoor air quality. Even the lower volume rate, size-selective samplers which have replaced the Hi-Vol in the U.S. may be too noisy for indoor use without modification. Other types of equipment designed for outdoor sampling of gaseous pollutants are noisy and often require sound-proof enclosures for use indoors. Samplers based on the chemiluminescence principle exhaust a certain amount of ozone which should be removed or exhausted outside when they are used indoors.

The degree of intrusion of an occupied structure that can be permitted is an important feature in the design of an indoor air quality monitoring program. In addition to the problem of noise, one must be concerned about the effect of the samplers on the normal activities of the occupants if these activities have an important bearing on the results. This is an especially important consideration in the use of personal monitors for acquiring total exposure data. Large, cumbersome samplers may change normal activity patterns producing bias in the data. In conducting fixed-point sampling in an occupied dwelling the location of sensitive monitoring equipment indoors may produce a potential problem with tampering and loss of data. One approach that has been used by many workers in the field is to mount all sampling equipment in an environmentally controlled trailer located outside the structure and drawing samples from within the structure to the samplers through long, inert, insulated tubes. However, tests must be carried out to assure that concentrations of pollutants are not affected by transport through long tubes.

Selection of the optimum monitoring configuration requires much thought and planning. This issue will be expanded upon in Chapter 2.

Five basic indoor air monitoring approaches are discussed below; all approaches are not available for some of the pollutants.

- *Continuous monitors* provide real-time monitoring with the capability of measuring short-term fluctuation and calculating long-term averages. Samplers must usually be operated at fixed locations either inside or outside the structure with sampling lines run to the sampling points. Through the use of sampling tubes and manifolds and suitable valving and electronic control, such devices can be operated as multipoint samplers. This technique is especially useful in characterizing indoor/outdoor air quality relationships and indoor air quality patterns. This method was first used in the U.S. by Yocom *et al.* (1971) and has subsequently been used by other researchers.

- *Dynamic time integrated samplers* collect pollutants on adsorbents, absorbents, filters, or impaction plates over fixed periods representing a fixed volume of air sampled. Such devices, which are usually cheaper than continuous monitors, can be mounted together in a sequential sampling device to collect a series of time averaged samples over an extended period. Such samples usually require chemical or physical analysis subsequent to sample collection.

- *Passive time integrated samplers* do not require a pump to draw air through a collection medium. Rather, the sensing of the pollutant is accomplished by the physical characteristics and dimensions of the sampler, the physical properties of the pollutants and diffusion of the pollutant to a chemically or physically sensitive surface or sector of the sampling device. Examples of such devices are the diffusion tubes for NO_2 monitoring developed by Palmes *et al.* (1976) and the Track-Etch[TM] samplers (Alter et al., 1981) used for measuring total alpha activity which is related to radon concentration.

- *Grab sampling* is used for indoor sampling when single, short-term samples can adequately describe the indoor pollutant concentrations. Such an approach would not normally be used when indoor air quality concentrations are varying rapidly unless a large number of grab samples can be taken at regular intervals. Examples of this method include filling evacuated flasks, cylinders, or inert plastic bags with indoor air samples for subsequent laboratory analysis, or using a hand pump to draw measured amounts of air over adsorbents in glass tubes.

- *Personal monitoring* devices are used to establish total exposure to a pollutant by individuals as they move about both indoors and outdoors. Such devices may be "dynamic" where air is drawn by battery-powered pumps through the sampling system or passive of the type described above. An example of the dynamic type is that used by the Harvard Six Cities Study (Dockery and Spengler, 1981). The Palmes tube (Palmes *et al.*, 1976) has also been widely used as a personal sampler for indoor NO_2 studies in relation to gas stove use. Wallace and Ott (1982) have conducted an in-depth review of personal monitors. The principal shortcoming of this method is that usually only integrated samples are collected. An important area of research is to develop personal monitors that will produce real-time results.

At present, there are no standardized methods for monitoring of indoor air quality, but a newly formed subcommittee (D22.05) of the American Society for Testing Materials (ASTM) is working aggressively on a program to standardize such methods.

1.3.2 Measurement of Air Exchange

Air inside of a structure exchanges with outdoor air as a result of infiltration or leakage and natural or mechanical ventilation. As can be seen from equations (1.1) and (1.2) air exchange has a profound effect on indoor air quality since infiltration of or ventilation with outside air increases the contribution of outside pollutants and dilutes those generated inside. Therefore, air exchange rate is an important variable to be measured as part of an indoor air quality study in order to be able to interpret the data. There are a number of approaches to the measurement of air exchange rate which are discussed below.

- *Indoor pollutant decay*. In some cases it is possible to use the decay rate of indoor generated pollutants as a means of calculating air exchange rate. One example would present itself in a study of the effect of unvented combustion devices in indoor air quality. The indoor air could be "spiked" with non-reactive combustion contaminants by turning on one or more devices until indoor concentrations reach an appreciable level. Then the devices are turned off and the concentrations of one or more of the unreactive combustion products (e.g., CO or CO_2) are tracked over time to produce an exponential decay curve. From a plot of the logarithm of concentration versus time the air exchange rate can be computed. Such a technique could also be used in an office building with significant CO_2 concentrations from human occupancy. At the end of the day when all workers and the sources of CO_2 leave, the decay of CO_2 with time can again be used to calculate air exchange rate. This technique can be used only when the source of the decay pollutants can be removed from the indoor space or turned off. Furthermore, only unreactive pollutants can be used in this way, and the concentration range during the decay period must be significantly above the outdoor concentrations.
- *Dynamic tracer tests*. Instead of using the decay of an existing pollutant to calculate air exchange, a unique gaseous compound can be introduced into the indoor space or into the ventilation system. Then, after adequate mixing of the tracer with the indoor air has taken place, the decay of this tracer gas can be used as before to calculate air exchange rate. The gas most commonly used as a tracer is sulfur hexafluoride (SF_6), a unique compound that is unreactive, non-toxic, and can be measured at extremely low concentrations using a gas chromatograph and electron-capture detector. This technique is used for obtaining short-term or essentially instantaneous air exchange rates. Sampling is usually carried out with grab samples using plastic bags, syringes, or evacuated tanks.
- *Passive tracer technique*. Dietz and Coté (1982) have developed an innovative tracer technique for determining air exchange rates using permeation tubes that emit known rates of perfluorocarbon tracers. Then passive monitors consisting of one or more activated charcoal tubes are strategically placed in the building to measure the integrated concentration of the tracer gases. With this method it is

possible to use emitting tubes with several different perfluorocarbons placed in different rooms making it possible to measure not only overall air exchange, but also room-to-room air exchange. Collected samples are analyzed for the tracer by gas chromatography, and the data are analyzed with the aid of a computer. Because of the miniscule quantities of perfluorocarbons emitted, it is necessary to expose the samplers over an extended period (several hours to a week), but the advantage of this method is that long-term air exchange rates are determined.

- *Fan pressurization technique.* A method for estimating air exchange rates indirectly is to use a "blower door." A calibrated fan is mounted in one of the doors to pressurize the structure after all other normal openings are sealed. Then the airflow is measured for a predetermined level of pressurization. This method does not measure air exchange rate directly. Rather, it gives an indication of the leakage area of the house. Persily (1982) has developed statistical models of the relationship between blower door results and measured air exchange rates.
- *Leakage area estimation.* This method is based on calculating air exchange rates for a building based on estimation of leakage areas for wall penetrations (e.g., doors and windows) and typical leakage data for construction components (e.g., walls and ceilings). Leakage rate data are available from the American Society of Heating Refrigerating and Air Conditioning Engineers (ASHRAE Fundamentals, 1985).

1.3.3 Measurement of Emissions

The measurement of emissions from indoor sources employs most of the same types of sampling equipment used in conducting indoor air quality monitoring studies. However, in measuring emissions there are several different configurations that may be used.

- *Hooding the source.* In this configuration a source is partially enclosed with a hood which is ventilated with a known flow rate of air. Samples for the emitted pollutants are collected in a duct between the hood and the ventilation fan. This configuration is commonly used to measure emissions from unvented combustion devices such as kerosene heaters. Measurement of the increase in pollutant concentration above background together with the flow rate in the ventilation system permits the calculation of pollutant emission rate. Precautions must be taken to assure that the hood captures 100 percent of the emissions and that the hood or any modified airflow patterns around the source do not affect the normal pollutant emitting characteristics of the source.
- *Chamber studies.* In this approach the source is placed in an environmental chamber with controlled exhaust or air exchange rate. After the source is allowed to emit pollutants at a normal rate and equilibrium is reached, in-chamber pollutant concentrations and air exchange rate are used to calculate pollutant emission rate. Materials of construction for the inside of the chamber walls

must be carefully selected to minimize surface reaction and/or deposition of the pollutants, and the dynamics of air movement in the chamber must be controlled to prevent pollutant decay or to achieve predictable rates of decay. A number of workers have used this technique to measure emissions from gas stoves and other combustion equipment (Coté *et al.*, 1974 and Girman *et al.*, 1982). Tichenor and Jackson (1986) at the U.S. EPA used a small chamber based on an incubator to measure emission rates of organic gases and vapors from household materials under a variety of environmental conditions.

- *Head space analysis*. This technique is in many ways similar to the chamber approach but on a much smaller scale. In one configuration small samples of an emitting material (e.g., a swatch of carpet or coated furniture fabric) are placed in a desiccator or other small, inert enclosure. Head-space gases are then collected and analyzed at atmospheric or reduced pressure and at ambient or elevated temperatures. Emission rate of a pollutant per unit quantity of material can then be calculated. Matthews et al. (1983) have developed a unique passive emission monitor to determine *in situ* emissions of formaldehyde from indoor surfaces.

1.4 References

Alter, H.W., Fleischer, R.L., Ginrich, J.E., and Murdock, S. (1981). "Passive Integrating Measurements of Radon and Thoron: Methods and Field Observations," in *Proceedings of the International Conference on Radiation Hazards in Mining*, M. Gomez (ed.), pp. 581–585. American Institute of Mining Engineers, New York, NY.

American Society of Heating, Refrigerating and Air Conditioning Engineers, Inc. *ASHRAE Handbook—Fundamentals, 1985*. ASHRAE, 1791 Tullie Circle, NE, Atlanta, GA U.S.A. 30329.

Biersteker, K., de Graff, H., and Nass, C. (1965). "Indoor Air Pollution in Rotterdam Homes," *International J. Air and Water Pollution* 9:343.

Coté, W.A., Wade, III, W.A., and Yocom, J.E. (1974). "A Study of Indoor Air Quality," EPA 650/4–74–042, U.S. Environmental Protection Agency, Washington, D.C.

Dietz, R.N., and Coté, E.A. (1982). "Air Infiltration Measurements in a Home Using Convenient Perfluorocarbon Tracer Technique," *Environ. Intl.* 8:419.

Dockery, D.W., and Spengler, J.D. (1981). "Personal Exposure to Respirable Particles and Sulfates," *J. Air Pollution Control Assoc.* 31:153.

Girman, J.R., Apte, M.G., Traynor, G.W., Allen, J.R., and Hollowell, C.D. (1982). "Pollutant Emission Rates from Indoor Combustion Appliances and Sidestream Tobacco Smoke," *Environ. Intl.* 8:213.

Matthews, J.G., Hawthorne, A.R., Daffron, C.R., and Reed, T.J. (1983). "Surface Emission Monitoring for Formaldehyde Source Strength Analysis," in *Proceedings Specialty Conference on Measurement and Monitoring of Non-Criteria (Toxic) Contaminants in Air*. Air Pollution Control Association, Pittsburgh, PA.

National Academy of Sciences (1981). *Indoor Pollutants*. National Academy Press, Washington, D.C.

Palmes, E.D., Gunnison, A.F., DiMattio, J., and Tomezyk, E. (1976). "Personal Sampler for NO_2," *Am. Ind. Hyg. Assoc., J.* 37:570.

Persily, A.K. (1982). "Understanding Air Infiltration in Homes," Ph.D. Thesis, Princeton

University, Center of Energy and Environmental Studies, Report PU/CEES no.129.

Tichenor, B., and Jackson, M. (1986). "The Measurement of Organic Emission from Indoor Materials—Small Chamber Studies," presented at the 1986 EPA/APCA Symposium on Measurement of Toxic Air Pollutants. Raleigh, NC, April 27–30, 1986.

U.S. Code of Federal Regulations (1989). 40 CFR, Part 50, U.S. Government Printing Office, Washington, D.C.

Wallace, L.A., and Ott, W.R. (1982). "Personal Monitors: A State-of-the-Art Survey," *J. Air Pollution Control Assoc.* **32**:601.

Yocom, J.E., Clink, W.L., and Coté, W.A. (1971). "Indoor/Outdoor Air Quality Relationships," *J. Air Pollution Control Assoc.* **21**:251.

Yocom, J.E. (1982). "Indoor–Outdoor Air Quality Relationships," *J. Air Pollution Control Assoc.* **32**:500.

CHAPTER 2

Planning an Indoor Air Quality Measurement Program

Indoor air quality measurements are made for a variety of purposes. This chapter outlines the principal types of indoor air quality studies and discusses some of the requirements of each type. One of the problems with the field of indoor air quality monitoring as it has developed is that many of the studies are anecdotal in character having been carried out to show the effect of some single or limited number of factors on indoor air quality in one or only a few indoor settings. The data from such programs may present data that accurately represent the effect of some factors that may influence indoor air quality (e.g., the effect of selected weatherization techniques on a few houses in a single geographical area). However, the limited scope of such studies together with monitoring approaches that may be different from those used in other similar studies make it difficult, if not impossible, to make comparisons of results between studies and to show some generalized effect considering all or a significant number of factors that influence indoor air quality.

Researchers in indoor air quality have tended to be less interested in developing data that may be comparable with data from other workers than in collecting data that demonstrate some specific effect. While the measurement methods used must provide accurate data, they have often been selected because of such practical constraints as manpower and equipment costs, and the availability of indoor environments rather than providing data that are comparable to data from other studies. As the indoor air quality field matures, it is hoped that more thought will be given to developing data bases that will have more widespread use than in the past.

There are two important reviews of protocols for monitoring indoor air quality. Nagda et al. (1987) have assembled useful information on various monitoring instruments available at the time their review was prepared. In addition, outlines for the design of indoor air quality monitoring programs are presented. Figures 2.1 and 2.2 are examples of generalized plans for an initial and detailed design, respectively.

The American Society for Testing Materials held a Symposium on "Design and Protocol for Monitoring Indoor Air Quality" in April 1987. The proceedings were published in 1989 (Nagda and Harper, 1989) and cover the subject in terms of three broad areas: Commercial and Office Buildings, Residential Buildings, and Instrumentation and Methods.

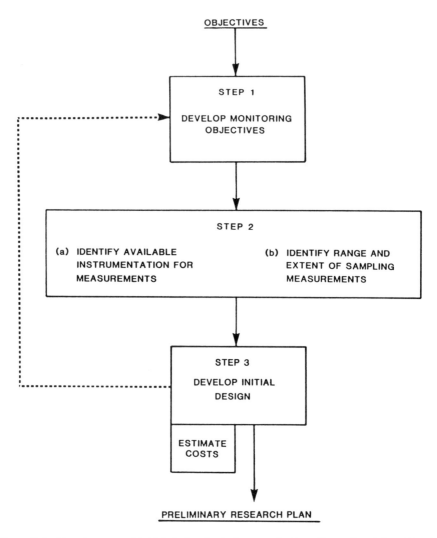

Figure 2.1   Development of initial design for indoor monitoring. (Reproduced from Nagda *et al.* (1987) with permission of Hemisphere Publishing Corp., New York)

Figure 2.2   Development of detailed design for indoor monitoring. (Reproduced from Nagda *et al.* (1987) with permission of Hemisphere Publishing Corp., New York)

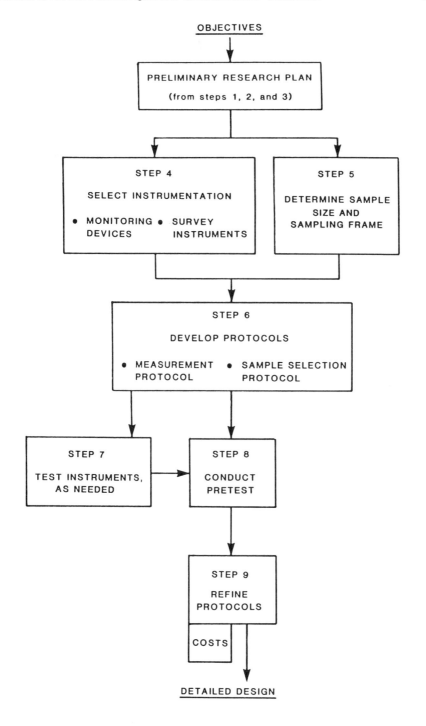

2.1 Establishment of Indoor Air Quality Profiles

One important type of indoor air quality study is to determine how indoor air quality varies in time and space and how indoor air quality compares with outdoor air quality. Such studies require the use of a number of samplers located at various points inside and outside a structure that are capable of measuring time-dependent concentrations either by sequential sampling or continuous recording instruments. One approach is to use a single continuous recording instrument fitted with multiple sampling lines running to several indoor locations with solenoid or rotary valves giving short sequential sampling periods for each sampling location. This method was used successfully by Yocom *et al*. (1971), Wade *et al*. (1975), Moschandreas *et al*. (1978), Ozkaynak *et al*. (1982), and Yarmac *et al*. (1987). Indoor air quality profiles can also be determined by use of individual samplers, but unless simple devices such as passive monitors are used, costs for such programs can be high.

In any indoor air quality monitoring program the question arises as to how uniform concentrations are within a given microenvironment. For naturally ventilated structures such as residences, the indoor atmosphere is usually relatively stable. Air movement is created by such factors as wind, infiltration, indoor–outdoor temperature differences, and opening and closing of windows and doors. In the absence of indoor sources, indoor concentrations of pollutants tend to be rather uniform. In the case of reactive outdoor pollutants penetrating indoors, concentrations tend to be high close to outside walls where infiltration occurs. Yocom *et al*. (1970) showed this effect for SO_2. Where there are indoor sources, strong pollutant concentration gradients can exist. An example is a home with an unvented gas stove or heater. In the case of a reactive pollutant such as NO_2, gradients are increased further because of the decay of the pollutant as it travels away from the source. A number of researchers have studied indoor gradients of NO_2 and CO in homes with unvented gas-fired appliances (Yocom *et al*., 1970; Wade *et al*. 1975, Moschandreas and Zabransky, 1982; Özkaynak *et al*., 1982; Yarmac *et al*., 1987). Here one finds extremely large variations in concentrations of combustion contaminants (NO_2, CO, and CO_2) in both time and space. For example, in homes with unvented gas stoves gradients as much as a factor of 2 or 3 can occur between the kitchen and bedroom depending on the air exchange rates and indoor airflow patterns. In studies such as these where there is concern about possible health effects from NO_2 and CO, it is necessary to define indoor human exposure to the pollutants of concern. This is commonly done by measuring concentrations simultaneously at multiple indoor locations together with determining occupant activity patterns during the monitoring period or by making measurements with personal monitors. Many types of personal monitors have the disadvantage of being unable to depict short-term, maximum exposures.

In the case of mechanically ventilated buildings, the variation of indoor pollutant concentrations is caused by a number of factors, including but not limited to:

- Variation in the source strength and location of indoor sources (e.g., smoking, use of office machinery, and occupancy patterns).
- Concentrations and variability of pollutants in the outdoor make-up air.
- Juxtaposition of supplies and exhausts for the ventilation system.
- Pattern for operating the ventilation system (e.g., differing amounts of make-up air depending on the season and weekend closedown).
- Patterns of airflow in the building and barriers to flow.
- Location of the air quality sampling points in relation to space dependent variables.
- Time period over which a sample is collected as it relates to time dependent variables.

In large mechanically ventilated office buildings spatial and temporal variations are generally not as extreme as in houses with strong indoor sources such as unvented combustion sources. However, the variation that does occur in mechanically ventilated buildings with good mixing is the result of spatially distributed indoor sources (smoking produces respirable particulate matter and volatile organic compounds (VOCs), human respiration produces CO_2, and furnishings and construction materials produce a wide range of VOCs). If the objective is not so much to define human exposure, but rather to determine as precisely as possible the effect of the relative amount of outdoor make-up air on indoor air quality, the monitoring protocol should include both fixed point monitoring and monitoring at several locations with portable instruments for certain of the key indoor pollutants and parameters.

Since carrying out a program in all parts of a large building would be extremely costly, great care should be taken to select those areas which will be most representative of occupant exposure or are selected to represent some other important factor such as pollutant sources or spatial gradients. The proper selection of sampling sites requires a detailed inspection of the building and its ventilation systems and may require the use of tracer studies as described in Chapter 3 to determine air distribution in relation to candidate sampling locations.

2.2 Exposure Monitoring with Personal Monitors

The simultaneous monitoring of indoor and outdoor concentrations of pollutants in and around a home provides data that can be used to infer total exposure to people that remain in or near the home (e.g., homemakers, preschool children, the aged and the infirm). However, for subjects that travel significant distances away from the home and travel through or occupy microenvironments with significantly different air quality than is found around the home, the only way to correctly assess total exposure to pollutants is to use personal monitors. These are devices that human subjects carry or wear and which continuously sample the air to which the subject is exposed as he or she travels through or occupies various microenvironments.

Some examples of major personal monitoring studies include the following:

- *Personal monitoring for CO*. A number of workers have used portable electrochemical CO monitors to conduct personal monitoring studies in a number of urban areas (Ott and Flachsbart, 1982; Ziskind *et al*., 1982; Akland *et al*., 1984; and Ott *et al*., 1986). These compact devices provide integrated or semicontinuous data on CO concentrations. Data on the subjects' activities are recorded on an activity log, or in the case of certain recently developed instruments, recorded in a data logger.
- *Personal monitoring for VOC*. The most important U.S. study of personal exposure to VOC is the Total Exposure Assessment Methodology (TEAM) Study (Wallace, 1987). In this study, human subjects carried Tenax GC^{TM} sorbent traps and small battery-powered pumps over two sequential 12-hour periods. Collected samples were analyzed by capillary gas chromatography/mass spectrometry (GC/MS).

Personal monitoring studies are designed to produce data on total human exposure to air pollutants which in turn may be used to assess health risk from such exposures. Therefore, the cohort of subjects used for such studies should be carefully selected to assure that they represent populations of interest. The selection process used in the TEAM study (Wallace, 1987) is an excellent example of a procedure to obtain an unbiased cohort of subjects.

Since personal monitoring represents an integration of exposure in a variety of microenvironments, subjects must keep activity logs to record periods spent in different microenvironments and in different activities. The TEAM study again can provide the model for the development and use of such logs.

Except for CO monitoring instruments which can provide a degree of real-time monitoring, most personal monitoring systems (e.g., the TEAM Study System) can only provide pollutant concentration data integrated over extended periods. This is an inherent disadvantage of most personal monitoring systems which will doubtless be rectified gradually in the future as more research is carried out.

2.3 Determining the Health Risk of Indoor Pollutants

Research on indoor air quality has often been conducted in conjunction with research on the health effects of air pollution. It is beyond the scope of this presentation to discuss the planning issues related to performing health effects studies. Such studies are truly multidisciplinary, involving epidemiologists, medical experts in the particular outcome being measured (e.g., respiratory or cardiovascular experts), statisticians, experts in survey research if information from questionnaires or diaries is to be collected, as well as experts in air monitoring. The range of health effects of interest depend on the pollutant(s) and the exposed population.

For example, research on CO has looked at the effect of personal exposure and ischemic heart disease in adults (Colome *et al.*, 1987) whereas research on NO_2 has looked at respiratory symptoms and pulmonary function in children (Vedal, 1985; Speizer *et al.*, 1980; Ware *et al.*, 1984).

The objectives of studies relating health effects to air pollutant exposure are massive in terms of the resources, personnel, and number of subjects in the study. This is necessary because the physiological changes being measured are variable and require a large number of subjects to detect a significant result. The magnitude of the study also depends on the study type, retrospective or prospective. Retrospective studies categorize exposure based on the presence or absence of a source of exposure (e.g., gas stove or cigarette smoker in the home). Prospective studies generally use indoor or personal air quality measurements to quantify exposure.

It is not possible to generalize about the significance of the health effects (e.g., to answer the question "does an indoor exposure cause an adverse health effect?") of indoor exposures because of the variation of pollutants, health outcome measured, and the quality of the study conducted. Different studies on the same pollutant have produced equivocal findings (e.g., Vedal, 1985). In discussing the concept of what defines an adverse health effect, Higgins (1983) succinctly divides the issue into four components:

- Is a certain biological change attributable to a specific compound?
- If so, what is the relationship between exposure and effect?
- Do persons with the change experience a higher morbidity or mortality than those without it?
- In the absence of future exposure, will the risk of morbidity or mortality increase?

By this decomposition of the issue of what is an adverse health effect, it is clear that the objective of the air monitoring study is to determine exposure in a manner that can be related to the anticipated health effect. For example, if acute health effects are of interest, the monitoring technique should measure peak exposures. On the other hand, if chronic health effects are being reported, integrated techniques should be used. Air monitoring specialists are critical members of interdisciplinary teams conducting health effects studies.

2.4 Approaches to "Sick Building" Studies

Much has been written about the "Sick Building Syndrome" (SBS). It is generally characterized by complaints from the building occupants about any or all of the following (Yocom and McCarthy, 1986):

- Building odor.
- Eye, nose, and throat irritation.
- Skin irritation.
- Feelings of lethargy.
- Feeling faint.
- Irritability.

The SBS problem is having an increasing impact in developed countries because of absenteeism and/or reduced efficiency of the workforce through concern about exposures to air contaminants in the workplace. The problem is exacerbated by a number of factors such as:

- Energy conservation in building heating and cooling.
- Energy conservation in renovated buildings.
- Increased use of synthetic building materials and furnishings.
- Increased public awareness of indoor air quality and the SBS.
- Changing attitudes among non-smokers about smoking in the workplace.

The presence of airborne contaminants is probably the most important feature of the building's interior that produces the undesirable symptoms, but any one of a number of other factors may also have a strong influence, such as:

- Temperature, humidity, movement, and mixing of air.
- Lighting.
- Sensitivity of the occupants to irritation.
- Changes in work location and crowding.
- Attitudes about the job and working conditions.

Nevertheless, it is often the buildup of air contaminants brought about by inadequate amounts of fresh air ventilation and its improper distribution that trigger the SBS.

In planning a study of an SBS situation, it is important to recognize the many factors that potentially contribute to the problem. The building owner or office managers in requesting help from an indoor air quality consultant to solve an SBS problem invariably ask for air sampling "to tell us what is in the air that could cause the problem." However, the problem should be approached in a series of sequential phases with the simplest and most cost-effective tasks performed first. Collection of samples requiring sophisticated and expensive analysis (e.g., GC/MS), should be done only if indicated by the results of earlier tasks. Table 2.1 is the outline of a simple protocol that has been used by the authors in the conduct of SBS studies. Such an outline is always discussed with the persons requesting SBS studies, and from this discussion a protocol specific to the problem at hand is developed.

Figures 2.3, 2.4, and 2.5 from the work of Woods *et al.* (1989) is a much more

Table 2.1 Outline of a survey protocol for sick building syndrome studies

A. *Initial Survey*
  (1) Discuss nature of the problems with building owners/operators, building engineers, and mechanical maintenance personnel.
  (2) Discuss nature of symptoms with complainants.
  (3) Initial Measurement Survey:
    (a) Measure CO_2 concentrations, relative humidity and temperature in various locations.
    (b) Conduct initial source survey. This might include contacting manufacturers of office supplies, building materials, furnishings, and cleaning products that are potential sources of indoor pollution.
    (c) Determine locations and sizes of air supply and return grills. Measure airflows in each, compare with ASHRAE standards and check for possible short-circuiting of air supply and return.
    (d) Conduct occupancy survey. Where are people located with respect to sources, air supply, etc.?
    (e) Determine location of main air intake and exhaust ducts on grills and assess possible recirculation and the conditions which might cause recirculation.
    (f) Review drawings and operating specifications for ventilation system and inspect the system for comparison with drawings and specifications. Inspect ducts for collection of contaminants (e.g., fiberglass) and inspect wetted surfaces and water reservoirs in air conditioning systems for growth of microbiological organisms.

B. *Detailed Questionnaire*
  Design and administer a questionnaire for building occupants or a portion thereof. The purpose of the questionnaire should be to objectively determine if the source of complaints can be correlated with a specific variable, e.g., work location, time of day or season, or sensitivity of the individual.

C. *More Detailed Site Survey (Scope based on results of A and B)*
  (1) Measure air exchange and mixing rates based on flow calculations and tracer decay tests.
  (2) Conduct indoor air quality and environmental measurements (measurements to be made in complaint and non-complaint areas as well as outdoors for comparison).
    (a) More measurements of CO_2, relative humidity, and temperature.
    (b) Sampling for individual contaminants, if potential sources are identified, such as formaldehyde, fibers, ammonia, and tobacco smoke components.
    (c) Broad spectrum analysis of organic gases and vapors using adsorbents analyzed by GC and GC/MS.
    (d) Sampling for biologically active contaminants (spores, molds or bacteria) in air and sampling of deposits in reservoirs and on wetted surfaces.

D. *Data Analysis*
  Assemble, analyze and interpret data collected in above phases. Compare results with any applicable standards (e.g., ASHRAE) or any data in the literature that show the relationship between exposure to an identified contaminant and human response. Air sampling and environmental monitoring data should be compared temporally and spatially with the results from the occupational questionnaire.

E. *Recommend Remediation Measures*
  Provide system designs and supervise installation of indoor air quality control or ventilation systems. Remind building owners/operators and office managers to inform occupants about any remediation measures that have been undertaken.

F. *Repeat applicable portions of the surveys under A, B, and C* to determine the effect of remediation measures on reductions in air contaminant concentrations and in the extent of complaints among building occupants.

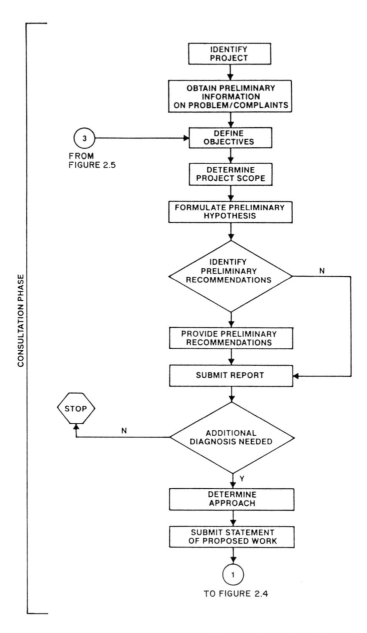

Figure 2.3   Flowchart for consultation phase of indoor air quality diagnostics protocol.
(Reproduced from Woods *et al*. (1989) with permission of the American Society for Testing
and Materials)

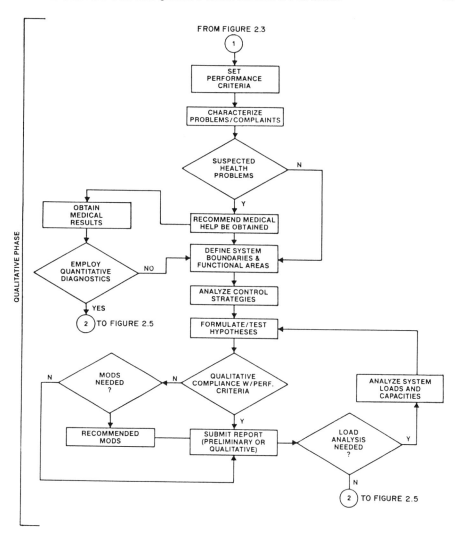

Figure 2.4 Flowchart for qualitative diagnostics phase of indoor air quality protocol. (Reproduced from Woods *et al.* (1989) with permission of the American Society for Testing and Materials)

elaborate protocol in the form of flow diagrams for the "consultation," "qualitative," and "quantitative" phases, respectively, of an SBS indoor air quality protocol. These flow diagrams provide enough detail for the indoor air quality researcher to be able to identify the key decision points and to use the project elements as a checklist for the design of an SBS study program.

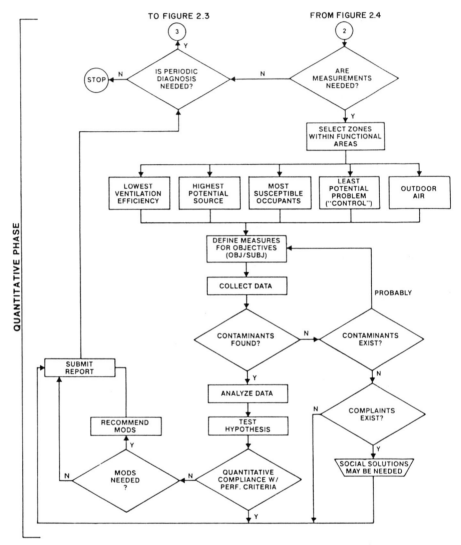

Figure 2.5   Flowchart for quantitative diagnostics phase of indoor air quality diagnostics. (Reproduced from Woods *et al*. (1989) with permission of the American Society for Testing and Materials)

Use of a properly designed questionnaire to obtain data on perceptions of building occupants to indoor air quality and climate is sometimes a helpful adjunct to an SBS study. However, at present there are no standardized questionnaire formats and methods of administration. Thus the results of questionnaire surveys of building occupants have been mixed. As more SBS and other indoor air quality studies are performed, progress in this area is anticipated. Wallace (1989) prepared a survey form used in the study of several government buildings. Lebowitz *et al*. (1989) published a questionnaire form developed by the Indoor Air Quality Committee of the Air and Waste Management Association. While this questionnaire was developed for use primarily in private residences, the rationale for its development and some aspects of the form may be applicable to SBS surveys.

2.5 Selection of Sampling Approaches

Indoor air quality and total exposure studies require three principal sampling modes:

- Sampling at fixed locations over significant lengths of time. The primary purpose is to determine air quality continuously or on an integrated basis at single points over some specified time period.
- Sampling for relatively short periods at a number of indoor locations (usually sequentially) where the sampler must be moved between samples.
- Personal monitoring in which the sampler is carried by or attached to a human subject whose exposure to a pollutant is being determined in various micro-environments.

2.5.1 Continuous Analyzers

Continuous air quality analyzers based on chemical and physical detection principles have been used for the measurement of gaseous criteria pollutants (CO, SO_2, NO_x, and O_3). These methods are well established and have been widely used to measure outdoor air quality to determine compliance with federal and state outdoor ambient air quality standards. These analyzers have reached a high degree of specificity, accuracy, and dependability. Many such systems have been used for indoor air quality investigations (especially NO_x and CO). Reliable continuous analyzers are also available for CO_2 and hydrocarbon gases and vapors, but thus far hydrocarbon analyzers lack adequate sensitivity for many indoor settings. A continuous analyzer for formaldehyde is also available, but it is not as reliable as the other instruments mentioned.

Advantages

- These analyzers can be operated unattended over extended periods.
- A continuous record of pollutant concentration is produced.
- Continuous analyzers respond to and record short-term peak concentrations.
- Results from sampling are usually immediately available.
- Output from samplers can be fed to data loggers.
- These analyzers can be "multiplexed" in which purged sampling lines from

several indoor (and outdoor) locations together with timer controlled valves can be used to collect semi-continuous sequential data from several locations. A further advantage of this system is that one instrument and one calibration can assure more comparable data from one location to another than if multiple instruments are used. In addition, such a "multiplexed" system is far more cost-effective than the use of multiple instruments.

Disadvantages

- Most continuous analyzers require an external power source and usually must be disconnected and reconnected if moved, and then must be recalibrated.
- Their size, complexity and requirement for calibration equipment and gases make them difficult to move. Furthermore, this characteristic may make such instruments difficult to place in confined indoor spaces which may influence normal indoor activity patterns.
- Such instruments may produce objectionable noise.
- Some instruments (e.g., those based on chemiluminescence) emit by-product gases that may need to be vented.

2.5.2 Integrated Dynamic Grab Samplers

This type of sampler takes many forms and can be used for a wide variety of pollutants. These samplers are called "dynamic" or "active" because sample air is mechanically or actively drawn through the device. This is in contrast to passive samplers which are discussed later. There are several basic concepts common to all samplers in this active sampling category: A known volume of sample air is introduced into a sample collection device over a known sampling interval; the collection device removes the pollutant(s) being collected by some chemical or physical principle; the quantity of pollutant collected is determined by appropriate chemical or physical laboratory procedures; and the average concentration of the pollutant is calculated for the sampling interval based on the quantity of pollutant measured and the air volume sampled.

Some types of samplers in this category can be configured to serve in the fixed point, portable, and personal monitoring modes. Although not exhaustive, the following list indicates the types of configurations possible for various classes of pollutants:

- Measurement of particulate matter concentrations by filtration and impaction with or without size discrimination.
- Collection and measurement of biologically active aerosols by impaction or liquid impingement.
- Collection of volatile and semi-volatile organic compounds (VOCs and SVOCs) on solid sorbents for subsequent separation by gas chromatography (GC) and

detection and/or analysis by flame ionization detection (FID), electron capture detection (ECD) and mass spectrometry (MS).

Advantages

- There is considerable versatility considering the large number of configurations and methods of pollutant collection available.
- In many cases the sensitivity of the method may be increased by increasing sample flow or sampling time. This kind of flexibility is important for methods developed for industrial occupational settings where concentrations of the pollutant being measured tend to be higher than in typical office buildings.
- Multiple units can often be combined in a sequential sampler that can produce a series of timed samples over a significant period of time.
- Sampling can be conducted at locations which may be inaccessible to continuous instruments.

Disadvantages

- The set up and analytical times for many of these methods are significant. However, in most such situations, the active grab sampling technique is the only feasible approach.
- These methods tend to require manual operation which increases costs and the risk of human error. Quality assurance procedures for sampling and analysis are often complex and time consuming.
- Some sampling methods require the handling of corrosive or toxic absorbing solutions which may pose hazards in occupied indoor spaces.
- Because of the time required for analysis of samples, the period between sampling and receipt of the results can be significant.
- Handling and shipping of the collection medium may allow the sample to become contaminated. The use of field blanks can be used to compensate for this, but increases the analytical costs.

A variation of this sampling technique is the use of inert bags (e.g., Tedlar™) or evacuated or pressurized cylinders to collect integrated samples. Such samples are returned to the laboratory for analysis. Collecting air samples in flexible bags requires the use of an airtight chamber in which the bag is placed and which permits the evacuation of the air in the chamber surrounding the bag to be evacuated at a known rate. Cylinders such as the SUMMA™ passivated stainless steel cylinders are currently being used by U.S. EPA (Winberry et al., 1988) for VOC sampling. These cylinders can be evacuated for subatmospheric pressure or pressurized sampling. In the case of bag or cylinder sampling, this approach is feasible only for pollutants and pollutant mixtures that are sufficiently unreactive or do not adsorb or condense on the walls of the container during the period between sampling and analysis.

2.5.3 Active Portable and Personal Monitors

There is an imperfect line of distinction between portable and personal monitors which is the reason for discussing them together. Conceptually, a portable indoor air quality monitor is an active monitor that can be readily moved from place to place in an indoor space to obtain a number of air quality samples in a reasonably rapid sequence in different locations. Most such monitors have battery-operated pumps. An active personal monitor is a portable monitor that can be attached to or be carried by a human subject without interfering with the subject's normal activities and produce a continuous record of total exposure of the subject to a given pollutant. Thus the distinction between a portable and personal monitor is related more to size and the degree of interference to human activity than any fundamental aspect of its operating principle. Some portable and personal monitors are capable of giving a direct readout of pollutant concentration while others are able to produce only an integrated reading for the operational period.

It is difficult to generalize about the advantages and disadvantages of portable and personal samplers because of the wide range of pollutants and operating principles, but listed below are a few of the advantages and disadvantages that hold in general for this class of samplers.

Advantages

• The small size of these monitors permits easy transport and unobtrusive sampling.
• Quiet operation is not a problem.
• These monitors record actual human exposure, not concentrations in a micro-environment.

Disadvantages

• While these systems can be used for fixed point sampling, battery charge limits the sampling period.
• Most samplers give only integrated readings. A notable exception is for CO monitoring where direct reading and data logging capability is available.

2.5.4 Passive Monitors

This is a unique category of instruments which has developed rapidly over the past few years. In this method the air reaches the surface through diffusion where the pollutant is captured by a chemically sensitive surface. Some are in the form of badges with a sensitive surface while others are tubes closed at one end with the sensitive surface on the inside of the closed end. Others are canisters which contain the trapping medium. A well known passive monitor is the one developed by Palmes *et al*. (1976) for the passive monitoring of NO_2. Originally developed as an industrial hygiene sampler, it must be exposed for several days in typical non-occupational indoor environments to give adequate sampling accuracy. Subsequent work by a number of other workers has increased the sensitivity of the method. Passive monitors are also available for HCHO, VOC and radon, and other types of integrating (passive) monitors are under development.

Advantages

- Passive monitors are simple to operate with no moving parts.
- These instruments are relatively inexpensive (purchase and analytical costs) as long as they are used in quantity.
- They can be used by inexperienced people (e.g., occupants of buildings) and are suitable for mail out monitoring programs.

Disadvantages

- They can produce only integrated concentration values.
- Exposure times tend to be rather long (days rather than hours).
- Some systems experience interference problems from other pollutants.
- Results may be affected by air currents and temperature.

2.6 Data Analysis

The approach to data analysis varies according to the nature of the investigation. In general, different approaches are used for the analysis of residential survey data and sick building investigations. This is due largely to study design and the number of samples collected. Residential research often generates more data than other types of studies and an attempt is made to control confounding variables. Sick building investigations usually produce a more limited amount of air quality data and confounding variables can usually not be controlled. In all cases the data analyses should be tailored to the objectives of the study. The following discussion outlines approaches to the analyses of data from both types of investigations.

2.6.1 Residential Research

Consultation of the indoor air quality literature reveals that researchers have used a wide variety of statistical techniques, from simple univariate measures to complex multivariate models. In the abstract, it is not possible to describe particular statistical techniques which should be used because analyses must be tailored to meet the objectives of the research. However, general guidelines can be described.

The first step is to become intimately familiar with the data set through various descriptive analyses, both univariate and bivariate. Although these analyses are relatively simple (and therefore often not critically reviewed), the importance of understanding the data at this level cannot be overemphasized. The interpretation of subsequent analyses will be enhanced by a thorough understanding of the data at this level. It is assumed that prior to doing any analyses, appropriate quality control/quality assurance procedures have been followed and any suspect data clearly identified.

Univariate analyses such as means, variances, percentiles, and probability density functions (pdf) (of which a frequency distribution is one type) should be evaluated. This last analysis will provide information on the shape of the distribution of the data, e.g., normal, log normal, and whether the pdf for all variables are similar. It is important to evaluate points at the tails of the distribution to identify if a pattern exists, e.g., temporally or spatially. The researcher should determine if these data make sense in light of other variables and the physical nature of the problem being investigated.

After conducting the univariate analyses, bivariate relationships in the data should be investigated. These analyses include: plotting variables against one another, time series plots, and a correlation matrix.

The second step in data analysis is often referred to as estimation techniques. In this step, insights gained from step 1 are used to guide the formulation of mathematical or statistical models or statistically testable hypotheses. Use of either of these techniques should be in accordance with the objectives of the study. The statistical models may be simple linear regression models or more complex multiple regression models, where the coefficients for the dependent variables are estimated from the data by a least-square minimization procedure. The goal is to have a set of dependent variables which are uncorrelated. These models are used to determine which dependent variables are important predictors of the independent (or outcome) variable and the amount of variance in the independent variable which is explained by the statistical expression.

Statistical inference methods require that the hypotheses, pertinent to the study objectives, be formulated in a statistically testable fashion. For example, if a researcher were investigating the influence of environmental tobacco smoke on indoor particle levels between homes with and without smokers present, the null hypothesis would be expressed as H_0: $X_1 - X_2 = 0$, where X_1 and X_2 are the sample means of the two different sets of homes. This hypothesis can then be tested using the t-test. If more than two groups of means are to be compared, then the appropriate test is an analysis of variance. Hypotheses can be nested for a series of analyses to address a more refined or detailed objective.

When conducting tests of statistical inference, the investigator must be cautious of what are called Type I and Type II errors. The source of these errors is the sample size. Type I error occurs when there is a significant difference between two samples, but it is not detected because the sample size is too small. Type II error occurs when there is no difference between two samples, but a statistical difference is observed because the sample size is very large. These errors can be accounted for in the experimental design phase (see any standard research design test or for a short discussion with application to an indoor air study, e.g. Nagda *et al*., 1987).

Caution should be exercised when interpreting statistical results from "exploratory" studies. As the description connotes, little may be known about critical variables which can influence the concentrations, and these variables are probably not controlled for in the experimental design.

Also included in this step is the development of a physical model, based on such relations as conservation of mass. An example of such a model is a mass balance model for indoor air quality which would include source and removal terms, re-entrainment, outdoor concentrations, and air exchange rate. The data analysis for the development of such models would include an evaluation of the pdf for the various pollutants to be included in the model, especially those having common source types. It may also be necessary to assess the significance of covariances among variables, e.g., when an occupant activity is a direct cause of the micro-environmental concentration.

The third step in data analysis is to evaluate model predictions, either statistical or physical. In this step model predictions are evaluated against observations. The data can be analyzed in a variety of ways including means, variances, and higher moments of the distribution, bias estimates between predicted and observed means and standard deviations, mean square error (MSE), and various goodness-of-fit criteria. The MSE can be decomposed for relative assessment of bias, and the residual (predicted minus observed) values should not be a function of any of the model variables. A physical model can be evaluated based on its ability to simulate the dynamic range of the data, stochastic variability, and confidence interval estimates. The final analysis of a physical model should be the comparison with new, independent data.

2.6.2 Sick Building Syndrome Investigations

This discussion focuses on the analysis of air quality data and not occupant survey data which may be collected during an SBS investigation. The reader is referred to Cox *et al*. (1988) for design issues and standard statistical tests on survey research for data analysis (e.g., Kerlinger, 1973).

Most SBS investigations tend to be conducted in a manner similar to detective work based on on-site observations and the history of complaints, as opposed to research plans based on the research design considerations discussed previously. The general guidance for SBS data analysis is to compare air quality results between suspect problem areas and control areas. This can be done either spatially (e.g., between two different areas of the building) or temporally (e.g., occupied versus non-occupied conditions) depending on the nature of the suspect source(s).

Funds for many SBS investigations are limited (by the client); thus the amount of data collected is small compared to large scale research projects on residential units. This means that data analyses are often qualitative as opposed to quantitative. If a sufficient number of samples has been collected, basic analysis such as t-tests or analysis of variance can be performed. These results can tell the investigator if concentrations are significantly different between two or more sampling conditions. More commonly insufficient data exist, and the investigator is required to make the best scientific/engineering judgement regarding an interpretation of the data based on previous experience and literature on the topic.

2.7 Conclusions

There is much that needs to be considered in designing an indoor air quality monitoring program. First and foremost is the purpose of the monitoring program. The approaches will depend on how the data will be used. Since field monitoring programs can be extremely expensive, it is important that careful planning together with pilot tests of monitoring procedures be carried out in advance of full scale field work. The amount of data needed and the required statistical analyses should be determined as part of the planning exercise.

2.8 References

Akland, G., Johnson, T., and Hartwell, T. (1984). "Results of the Carbon Monoxide Study in Washington, D.C., and Denver, Colorado, in the Winter of 1982–83," EPA Report No. EPA-600/D-87–178.

Colome, S.D., Lambert, W.E., and Castaneda, N. (1987). "Determinates of Carbon Monoxide Exposure in Residences of Ischemic Heart Disease," *Proceedings of the 4th International Conference on Indoor Air Quality and Climate*, Institute for Water, Soil and Air Hygiene, Berlin.

Cox, B.G., Mage, D.T., and Immerman, F.W. (1988). "Sample Design Considerations for Indoor Air Exposure Surveys," *J. Air Pollution Control Assoc.* **38**:1266–1270.

Higgins, I.T. (1983). "What is an Adverse Health Effect?," *J. Air Pollution Control Assoc.* 7:661–663.

Kerlinger, F.N. (1973). *Foundations of Behavioral Research*, 2nd edn, Holt, Rinehart & Winston, New York.

Lebowitz, M.D., Quackenboss, J.J., Soczek, M.L., Kollander, M., and Colome, S. (1989). "The New Standard Environmental Inventory Questionnaire for Estimation of Indoor Concentrations," *J. Air and Waste Management Assoc.* 25:1411–1415.

Moschandreas, D.J., Stark, J., McFadden, J.E., and Morse, S.S. (1978). "Indoor Air Pollution in the Residential Environment," Geomet Report, Final Report for the U.S. EPA, Report No. 600/7–78/229, Vols I and II, U.S. Environmental Protection Agency, Washington, D.C.

Moschandreas, D., and Zabransky, J. (1982). "Spatial Variation of Carbon Monoxide and Oxides of Nitrogen Concentrations Inside Residences," *Environ. International* **8**:177–183.

Nagda, N.L., Rector, H.E., and Koontz, M.D. (1987). *Guidelines for Monitoring Indoor Air Quality*, Hemisphere Publishing Corporation, New York, NY.

Nagda, N.L., and Harper, J.P. (eds). (1989). *Design and Protocol for Monitoring Indoor Air Quality*, STP 1002, American Society for Testing Materials, Philadelphia, PA.

Ott, W.R., and Flachsbart, P. (1982). "Measurement of Carbon Monoxide Concentrations in Indoor and Outdoor Locations Using Personal Exposure Monitors," *Environ. International* **8**:295–304.

Ott, W.R., Rodes, C.E., Drago, R.J., Williams, C., and Burman, F.J. (1986). "Automated Data-Logging Personal Exposure Monitors for Carbon Monoxide," *J. Air Poll. Control Assoc.* **36**(8):883–887.

Ozkaynak, H., Ryan, P.B., Allen, G.A., and Turner, W.A. (1982). "Indoor Air Quality Modeling: Compartmental Approach with Reactive Chemistry," *Environ. International* **8**:461–471.

Palmes, E.D., Gunnison, A.F., DiMattio, and Tonsczyk, C. (1976). "Personal Sampler for Nitrogen Dioxide," *Am. Ind. Hyg. Assoc.* **37**:570.

Speizer, F.E., Ferris, B.G., Jr., Bishop, Y.M.M., and Spengler, J.D. (1980). "Respiratory Disease Rates and Pulmonary Function in Children Associated with NO_2 Exposure," *Am. Rev. Resp. Dis.* **121**:3–10.

Vedal, S. (1985). "Epidemiological Studies of Childhood Illness and Pulmonary Function Associated with Gas Stove Use," in *Indoor Air and Human Health*, Lewis Publishers, Inc., Chelsea, Michigan.

Wade, W.A., Coté, W.A., and Yocom, J.E. (1975). "A Study of Indoor Air Quality," *J. Air Pollution Control Assoc.* **25**:933–939.

Wallace, L.A. (1987). *The Total Exposure Assessment Methodology (TEAM) Study*, EPA/600/6–87/002a, U.S. Environmental Protection Agency, Washington, D.C.

Wallace, L.A. (1989). Personal Communication to S.M. McCarthy.

Ware, J.H., Dockery, D.W., and Spiro, A. (1984). "Passive Smoking, Gas Cooking and Respiratory Health of Children Living in Six Cities," *Am. Rev. Resp. Dis.* **129**:366–374.

Winberry, W.T., Jr. Murphy, N.T., and Corouna, B. (1988). "Compendium of Methods for Determination of Air Pollutants in Indoor Air," Contract No. 68–02–4467, U.S. Environmental Protection Agency, Research Triangle Park, NC.

Woods, J.E., Morey, P.R., and Rask, D.R. (1989). "Indoor Air Quality Diagnostics: Qualitative and Quantitative Procedures to Improve Environmental Conditions," in *Design and Protocol for Monitoring Indoor Air Quality*, Nagda, N.L. and Harper, J.P., eds, STP 1002, American Society for Testing and Materials, Philadelphia, PA.

Yarmac, R.F., McCarthy, S.M., and Yocom, J.E. (1987). "Final Report on an Unvented Gas Space Heater Study," prepared for the Consumer Product Safety Commission, Washington, D.C.

Yocom, J.E., Clink, W.L., and Coté, W.A. (1970). "A Study of Indoor-Outdoor Air Pollution Relationships, Vol I," Final Report prepared for the National Air Pollution Control Association, Washington, D.C.

Yocom, J.E., Clink, W.L., and Coté, W.A. (1971). "Indoor/Outdoor Air Quality Relationships," *J. Air Poll. Control Assoc.* **21**:251–259.

Yocom, J.E., and McCarthy, S.M. (1986). "Indoor Air Quality: The Tight Building Syndrome," *Building Operating Management Magazine* **33**: No. 2, 82–88, Milwaukee, WI.

Ziskind, R., Fite, K., and Mage, D. (1982). "Pilot Field Study: Carbon Monoxide Exposure Monitoring in the General Population," *Environ. International* **8**:283–293.

CHAPTER 3

Building Dynamics: Theory and Measurement of Infiltration

It is evident from the preceding discussions that a dynamic relationship exists between the air within and the air external to a structure. Indoor air quality is affected by numerous parameters: indoor sources, physical and chemical removal processes, and the rate at which outdoor air enters and indoor air exits the structure. In this chapter, the theoretical and empirical aspects of air infiltration and ventilation are presented.

Air moves in and out of buildings at varying rates depending upon a number of factors relating to both the structure and the local meteorological conditions. Two terms are used to describe how air enters a building: infiltration and ventilation. Both are measured as air exchange rate, or air changes per hour (ACH).

The American Society of Heating, Refrigerating, and Air Conditioning Engineers (ASHRAE, 1989) defines infiltration as "uncontrolled airflow through cracks and interstices, and other unintentional openings." Infiltration occurs because no building is completely airtight; wind pressures and temperature differences create driving forces which push or draw the outdoor air through openings into the building. Exfiltration is air that leaks out through these or other openings.

There are two types of ventilation, natural and mechanical. Both are distinct from infiltration. Natural ventilation refers to the intentional displacement of air through specified openings such as windows, doors, and vents without fans. Mechanical ventilation refers to the use of air movers (fans) to bring in outdoor air or to exhaust indoor air.

Infiltration is the rate of exchange of outdoor air with the entire volume of indoor air, quantitated as ACH. If a building has a total air exchange rate of 1 per hour, a volume equivalent to the total internal volume of air is replaced by outdoor air once per hour. Infiltration rates in residential structures may vary from less than 0.2 ACH, which would be considered low, to greater than 2.0 ACH. Infiltration rates in the high range tend to be characterized by noticeable drafts around windows, doors, and other openings in the building envelope. Figures 3.1 and 3.2 show respectively histograms of natural infiltration rates for new construction and low income housing in North America (ASHRAE, 1989). The median infiltration rate for new construction is 0.5 ACH, while that for low income housing is 0.9 ACH.

Air exchange between outdoors and indoors, in whatever form, dilutes

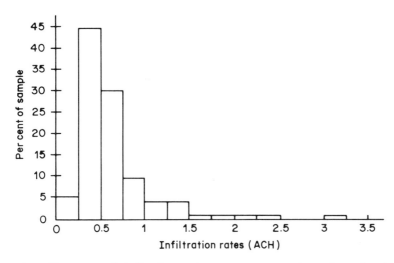

Figure 3.1   Histogram of infiltration values—new construction. (Reproduced from ASHRAE (1989) with permission of the American Society of Heating, Refrigerating and Air Conditioning Engineers)

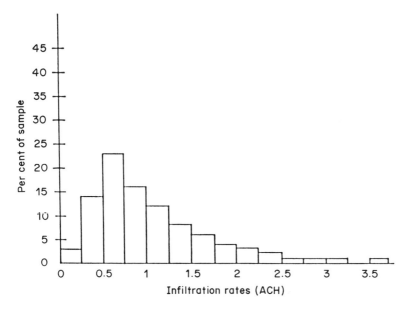

Figure 3.2   Histogram of infiltration values—low-income housing. (Reproduced from ASHRAE (1989) with permission of the American Society of Heating, Refrigerating and Air Conditioning Engineers)

contaminants generated indoors with outdoor air. However, when contaminant levels are higher outdoors than indoors, contaminant migration to the indoors will occur. In this chapter we present a discussion on the parameters which drive infiltration, their relationships, measurement techniques, and some examples of air exchange rate measurement data. The sections dealing with effects of temperature, wind, and building features are based on materials prepared by ASHRAE (1981, 1985, and 1989). Those wishing for further details on these topics and guidance in calculating infiltration are directed to these publications. Other references worth consulting are Liddament (1986) for theoretical calculations and Charlesworth (1988) for measurement techniques; both published by the Air Infiltration and Ventilation Centre in Great Britain.

3.1 Factors Affecting Air Infiltration

Four major factors affect the infiltration rates in homes: type of structure and construction, meteorology, heating and cooling systems, and occupant activity. Meteorology is the main driving force for infiltration and the basic building block of any model for infiltration rate estimation. The other factors influence infiltration but are difficult to quantitate in a theoretical manner. As discussed in the following paragraphs, infiltration is a dynamic parameter which is influenced by many factors with structural and meteorological conditions being the most important.

3.1.1 Structural Parameters

Infiltration rate is influenced by several structural factors, some of which are easier to characterize and quantitate than others. The first factor relates to the construction features of the building. This includes quality of construction, materials of construction and condition of the structure. Quality of construction refers to the size of the air gaps between building shell components. The fewer the gaps the lower the rate of infiltration. These gaps are built in during the construction phase and are usually difficult to modify after the fact. It is difficult to estimate the looseness or tightness of a home from the blueprints because infiltration depends on how the building is assembled. Materials of construction refers to the types of windows and doors, wall/ceiling details, heating system, wall framing details (e.g., balloon framing) and fireplaces used in the building. ASHRAE (*Fundamentals Handbook*, 1985) notes that both interior and exterior walls are important to consider; e.g., brick is leakier than a frame wall, and plastering interior walls effectively reduces infiltration through the walls to negligible levels. Materials of construction and quality are regionally dependent. The condition of the structure also affects infiltration rates. In general, infiltration increases as a house ages. Homes settle and new leaks can develop around the sill and at window and door frames; caulking and weatherstripping can deteriorate; and mortar may crack and wood warp. All of these items can be identified by inspection, but the magnitude of their influence on infiltration rate is difficult to measure.

3.1.2 Meteorological Parameters

The airflow rate due to infiltration, exfiltration, and natural ventilation depends upon pressure differences between the inside and outside of the structure and the resistance to flow through building openings. Pressure differences are caused by wind pressure in relation to the position of building openings and temperature differences between indoors and outdoors.

Wind Effects

The wind field surrounding a building provides a driving force to increase air infiltration. Positive pressure exists on the windward side and negative pressure exists on the leeward side of the building. The magnitude of the positive pressure on the windward side is sensitive to the orientation of the flat surfaces of the building. Further, the extent to which this pressure enhances infiltration is strongly dependent upon the number, size and location of air leakage paths.

Blomsterberg and Harrje (1979) emphasize the importance of the location of openings and leakage paths. They believe that the locations of building openings can produce a factor of two range in infiltration rates.

A number of factors influence the extent of wind effects on air infiltration and make it difficult to develop a rigorous relationship between wind speed and air infiltration. They include:

- Shell and exterior air barriers.
- Interior barriers to flow (e.g., walls, ceilings, and closed doors) that cause internal pressure buildup and thus reduce infiltration.
- Lack of precise knowledge of the detailed wind pressure profiles on building surfaces.
- Influence of complex terrain, presence of trees and other obstacles that create channeling and may increase the magnitude of wind force and alter its direction close to the structure.
- Sheltering, urban canyon and building wake phenomena due to surrounding buildings and other neighborhood factors.
- Fluctuating winds, rather than linear wind forces, that may affect infiltration rates through window cracks.

Temperature Effects

The temperature inside a structure is often different from the outside ambient temperature. Maximum temperature differences occur when the indoor environment is heated. Temperature differences cause differences in air density inside and outside, which in turn produce pressure differences. In the winter when indoor air temperatures are high relative to those outdoors, the warmer and less dense air inside rises and flows out of the building at its top. This air is replaced by cold outdoor air that enters near the bottom of the building or from the ground (i.e., soil gas). This phenomenon is called the building "stack effect." During hot weather when air conditioning produces lower temperatures inside than outside, the reverse process occurs. However, indoor–outdoor temperature and pressure differences during the summer are usually not as great as during the winter.

For a heated building the inside vertical pressure gradient relative to that outdoors is depicted in Figure 3.3 for buildings with cracks and other leakage paths evenly distributed along the height of the buildings and no internal airtight membranes. Note that there is a point along the vertical distance of the building where inside and outside pressures are equal. This is called the neutral pressure level (NPL). Under the assumed conditions of no internal partitions and evenly distributed leakage paths, the NPL is one half the building height. A useful rule of thumb for estimating the magnitude of the stack effect in a building is that the pressure difference induced by the stack effect is 2.7×10^{-5} in. water/FT °R (ASHRAE, 1989). This estimate neglects the effect of any resistance to vertical flow within the building.

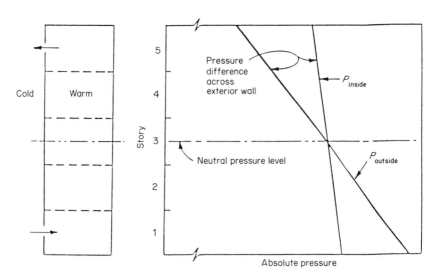

Figure 3.3 Stack effect in a building with no internal partition. (Reproduced from ASHRAE (1981) with permission of the American Society of Heating, Refrigerating and Air Conditioning Engineers)

If the building contains airtight membranes (floors and ceilings) that separate the stories, the pattern of pressures is modified. If each story is separated by a completely airtight membrane, each story would operate as an individual cell and infiltration from the stack effect would tend to be minimal. In real life, each story is not completely airtight and pressure patterns could be quite complex. For a detailed discussion of the quantitative effects of temperature differences on air infiltration in buildings of various configurations, the reader is referred to ASHRAE (1989).

An important point to be kept in mind is the effect of reduced pressures on the basement. Migration of soil gas in the region adjacent to the outside basement wall into the basement will be influenced by basement pressure. Soil gas may contain such contaminants as radon, pesticides and VOC from nearby landfills or leaking underground fuel storage tanks. As shown below, operation of a furnace in the basement can produce even lower basement pressure.

Humidity

Humidity is another factor which may affect infiltration but which has not yet been studied thoroughly. Luck and Nelson (1977) found a correlation between humidity levels and infiltration rates and hypothesized that swelling or shrinking of wood with changes in humidity can alter crack dimensions. Stricker (1975) reported that homes with low infiltration rates also had high humidity. It may be that humidity decreases infiltration and this in turn contributes to an increase in the indoor humidity. In a study by Yarmac *et al.* (1987) in 25 houses in the southern U.S., no apparent relationship was found between relative humidity and air exchange rate. One explanation for this lack of association is that absolute humidity, rather than relative humidity, may be a better measure of any effect the water content of the air has on infiltration. More studies are needed to characterize the interrelationship between infiltration and humidity.

3.1.3 Heating, Cooling and Exhaust Systems

In addition to meteorological variables, heating, cooling, and exhaust systems used in a home can affect the rate of air infiltration. The primary factors are the type of system and the duration of operation. Air, for the combustion of fossil fuels or wood, is extracted from the basement or living space. This interior air is then replaced through increased infiltration. The duration and frequency of furnace operation determines the relative importance of this source for infiltration.

Exhaust and forced air ventilation fans in a house will change indoor pressure dramatically. The fundamental effect will be to change the location of the NPL. If exhaust ventilation is provided to the building depicted in Figure 3.3, the indoor pressure distribution line will shift to the left and the NPL will rise. Conversely, if forced ventilation is provided, the indoor pressure distribution line will shift to the right and the NPL will be lowered. In a home with a furnace in the basement, the requirement for combustion air will act like an exhaust fan in raising the NPL. Figure 3.4 is a schematic diagram of a house with three stories and a basement and shows the effect of operating a furnace in winter on the position of the NPL. The vertical pressure distribution is shown for a typical situation during the winter with

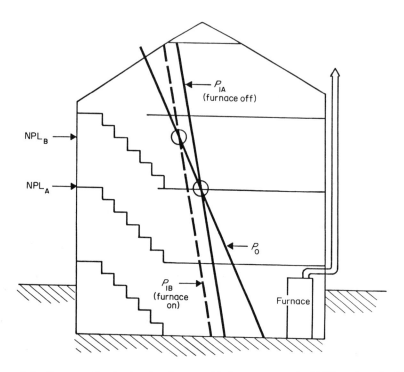

Figure 3.4   Pressure distribution in a house with furnace on and off (Yocom *et al.*, 1986)

the furnace on and off. When the furnace is on and the NPL rises, infiltration will occur over a larger area of the building and there will be greater negative pressures at the bottom of the building, especially in the basement.

The effects of furnace combustion air on air exchange was estimated for a house of $16\,000$ ft^3 (453 m^3) volume with a furnace using one gallon/hour (3.8 liter/h) of No. 2 fuel oil (Yocom *et al.*, 1986). Assuming 30 percent excess air, the air exchange rate would increase by 0.11 ACH during periods of furnace operation. Depending on how tightly the house is built, furnace operation may have a significant effect on the indoor–outdoor pressure differences and the air exchange rates.

During periods of furnace operation, negative basement pressures increase the penetration of soil gas surrounding the basement walls. However, if the primary infiltration leakage paths are in the upper stories of the house, airflow induced by combustion air requirements would tend to be towards the basement. Such air would presumably dilute to some degree any contaminants entering through the basement walls.

Comparing air infiltration in air conditioned spaces with infiltration in heated spaces is not straightforward since air conditioning usually includes mechanical ventilation. During the warmer months when such systems are in use in the recirculatory mode, the building is kept closed to reduce the loss of cool air. The house will have a low rate of infiltration relative to a house which has natural ventilation from the opening of windows. An air conditioned home will trap more indoor pollutants than a non-air conditioned home during such periods. Furthermore, when leaky duct work runs through a non-air conditioned portion of the home (e.g., crawl space) the operation of the cooling system has a dramatic effect on the air exchange rate.

3.1.4 Occupant Effect

In addition to all the physical parameters which affect infiltration rate, the activity of the occupants also has an influence. These activities include: opening and closing of external doors, operating kitchen and bathroom exhaust fans, operating heating systems and the opening and closing of interior doors, all of which will change the impact of interior barriers to airflow. Quantifying the relative importance of these effects on infiltration (or natural ventilation) rates is difficult because of large individual variation. Brundrett (1977) found that in Britain the window opening behavior was affected by the size of the family, if a resident was at home, and weather conditions. Lyberg (1981) conducted a survey of homes in Sweden and found that the fraction of homes with open windows was related to the indoor–outdoor temperature difference and windows on the leeward side of the building were the most frequently opened. Both the English and the Swedish studies found that the most frequently opened windows were in the bedroom. The English study noted that humidity control was the reason for opening windows in the winter. In a study of 50 California homes, Alevantis and Girman (1989) showed that when windows were open in July, a warm month in California, air exchange rates could increase to over 5 ACH.

3.2 Fundamental Relationships of Air Infiltration

As noted previously, infiltration and ventilation are driven by pressure differences. The most important causes of these pressure differences are indoor–outdoor temperature and wind pressure. Other items which affect pressure differences are combustion appliances and mechanical ventilation systems. The following discussion focuses on pressure differences caused by wind and temperature and their interrelationship.

The fundamental relationship is: mass flow of air into a building equals the mass flow of air out of the building. Assuming density differences are negligible, volumetric airflow entering a building equals the volumetric airflow leaving the building.

The simplest condition exists when there is no indoor–outdoor temperature difference and wind pressure is the only driving force (assuming no appliances are operating). The pressure difference across the building envelope is expressed as follows (ASHRAE, 1989):

$$\Delta P = P_0 + P_w - P_i \qquad (3.1)$$

where:

ΔP = pressure difference between outdoors and indoors at the location
P_0 = static pressure at reference height in the undisturbed flow
P_w = wind pressure at the location
P_i = interior pressure at the height of the location

The more usual case is when both wind and indoor–outdoor temperature differences contribute to the ΔP across the building envelope. Temperature differences impose a gradient in the pressure difference which is a function of height and the temperature difference. This effect is additive to the wind pressure expression and is expressed as (ASHRAE, 1989):

$$\Delta P = P_0 + P_w - P_{i,r} + \Delta P_s \qquad (3.2)$$

where:

ΔP_s = the pressure caused by the indoor–outdoor temperature difference (stack effect)
$P_{i,r}$ = the interior static pressure at a reference height ($P_{i,r}$ assumes a value such that inflow equals outflow)

It is not practical to calculate ΔP from these equations as it is not possible to measure the individual pressures (interior static, wind or temperature induced). To do this, measurements would have to be made around each opening in the building envelope. However, it is possible to relate P_w and ΔP_s to more easily measured parameters, wind speed and temperature.

The relationship between wind speed and wind pressure is given by Bernoulli's equation (ASHRAE, 1989):

$$P_v = \frac{C_p \rho V^2}{2} \qquad (3.3)$$

where:

P_v = surface pressure relative to static pressure in undisturbed flow, Pa
C_p = surface pressure coefficient
ρ = density of air, kg/m^3
V = wind speed, m/s

Under standard conditions (101.3 Pa or 14.7 psi) and 20 °C, this equation reduces to:

$$P_v = C_p 0.601 V^2$$

C_p varies with location around the building envelope and wind direction. See ASHRAE (1989) for further information and values for C_p.

The differences in air density due to temperature differences between the interior and the exterior of a building create the pressure difference which drives infiltration. To estimate this pressure difference, ΔP_s, it is necessary to know the NPL. This pressure difference can be expressed as:

$$\Delta P_s = \rho_i g h (T_i - T_o)/T_o \qquad (3.4)$$

where:

ΔP_s = pressure difference, Pa
ρ_i = density of air, kg/m^3
g = gravitational constant, 9.8 m/sec^2
h = distance to NPL (positive if above, and negative if below) from the location of the measurement

subscripts:

$_i$ = inside
$_o$ = outside

It is difficult to know the location of the NPL at any one moment, but there are some general guidelines. According to ASHRAE (1989), the NPL in tall buildings can vary from 0.3 to 0.7 of total building height. In houses with chimneys, it is usually above mid-height, and vented combustion sources used for space heating can move the NPL above the ceiling.

No general statement can be made on whether wind effects or stack (temperature) effects are more important to infiltration; this relationship is site-specific. It depends on such factors as building height, internal resistance to airflow, local terrain, orientation of the building relative to the prevailing wind, and shielding of the building. Because of this, there is no universal theoretical expression for combining wind and stack effects to calculate infiltration. ASHRAE (1989) presents one simplified approach:

$$Q_{ws} = (Q_w{}^2 + Q_s{}^2)^{0.5} \qquad (3.5)$$

where:

Q_{ws} = infiltration from both wind and stack effects
Q_w = infiltration from the wind
Q_s = infiltration from the stack

The foregoing discussion is meant to highlight how the wind effect and stack effect jointly create the pressure differences around openings in the building envelope which give rise to infiltration. The equations, as presented, are not intended to be used to calculate air exchange rates. ASHRAE (1981, 1985, and 1989) presents equations that express indoor–outdoor pressure differences in relation to wind speed and stack effect pressure differences. These same publications present tables of data on such factors as effective leakage areas for building components under the influence of some nominal pressure difference. By combining the equations with the data on building components, it is possible to calculate infiltration rates.

3.3 Measurement Techniques

The physical concept of air exchange is straightforward, yet the actual quantitation is dependent on the measurement technique which only approximates the theoretical model. This is because the methods of measurement focus on one parameter of the theoretical model. Other parameters are estimated.

This section presents an overview of three techniques for measuring infiltration: tracer gas, fan pressurization, and effective leakage area. The reader is encouraged to consult original references and researchers in the field as there are nuances, advantages and disadvantages for each technique which are too detailed to present here. A drawback of the tracer gas methods is that in most applications they are "snapshots" of the infiltration rate: one-time measurements made to characterize a structure under given conditions. However, air infiltration is a dynamic property which changes as the driving forces change. Empirical models exist for extrapolation of exchange rates to other conditions, but long-term measurements (e.g., seasonal) are not conducted frequently. Fan pressurization measurement is less influenced by prevailing meteorology than other methods, but it is an indirect measurement of actual infiltration. Airflow through leaks in the building envelope is measured at artificially induced pressure differences. The effective leakage area method, another indirect method, estimates air exchange by using an empirical model to predict air exchange rate from fan pressurization data.

In the selection of a measurement method for air exchange rate, several parameters must be taken under consideration. The most important one is choosing an air exchange rate method which complements other air quality data (in terms of averaging time) being collected and the goals of the research. For example, if continuous pollutant data are being collected over a relatively short time period, then the tracer gas method may be most appropriate. However, if integrated measurements are collected over long periods of time (e.g., more than one sea-son), then the fan pressurization or effective leakage area method may be appro-priate. Other considerations include type of data analysis being performed (e.g., exploratory or model building or verification) and the overall cost of the project.

3.3.1 Tracer Gas

Tracer gas decay is a direct measurement of air exchange rate. An inert gas which is easily detected at very low concentrations is released and uniformly mixed within the building. Assuming the gas does not react chemically or physically with the surrounding materials, the gas concentration will decrease as dilution air flows into the building. The rate of decrease is proportional to the infiltration rate. Ideally, successive measurements over extended periods of time are necessary to determine the relationship between the rate of air infiltration and meteorological conditions for a specific building.

There are three key assumptions of the tracer gas technique: the tracer gas mixes perfectly and instantaneously; the effective volume of the enclosure is known; and the factors that influence air infiltration remain unchanged throughout the measure-ment period (Liddament and Thompson, 1983). Imperfect mixing occurs when air movement is impeded by flow resistances or when air is trapped by the effects of stratification. This causes a spatial variation in the concentration of the tracer gas within the structure. Sampling locations may be biased by this effect. Fans are often used to mix the tracer gas with the building air. Effective volume is assumed to be the physical volume of the occupied space. Areas which contain dead spaces that do not communicate with the rest of the living space will reduce the effective volume. Variation in conditions during the measurement period, such as door openings or meteorological changes, will cause a departure from the logarithmic decay curve and the equation on which infiltration is calculated will no longer hold. Due to the influence of weather conditions it is preferable to have on-site measurements of wind speed and temperature.

There are several different types of gases used as tracers: helium, nitrous oxide, carbon dioxide, carbon monoxide, sulfur hexafluoride (SF_6), and perfluorocarbons (PF). These tracer gases share certain desirable characteristics; they are non-toxic at concentrations normally used in such studies, non-allergenic, inert, non-polar, and can be detected easily and at low concentrations. The gases which are used most frequently are SF_6 and perfluorocarbons. Carbon dioxide or carbon monoxide can be used if initial concentrations are substantially above background but well below concentrations of health concern.

While it is true that each of the tracers mentioned above are extremely unreactive, it is not certain that they do not adsorb on or desorb from indoor sources. The U.S. EPA is conducting chamber studies on SF_6 to determine if such effects occur and whether any such effects may tend to influence air exchange measurements using this tracer (Tucker, 1990). More research is clearly needed in this area.

Within the category of tracer gas methods for determining air exchange rate are two different application techniques. The first method, pulse release, involves the release of a known amount of gas, mixing it thoroughly within the space, and taking grab air samples at specified intervals to determine the decrease in concentration with time. Another approach is the constant release rate technique; the tracer gas is released at a known rate via a permeation tube and integrated samples are collected with absorption tubes, usually over one to several days. The air exchange rate is proportional to the amount of tracer gas which is collected, given the volume of the space and the release rate.

The grab sampling approach and the integrated sampling approach find the widest application in this field and are summarized below. As the grab sampling method most frequently uses SF_6 as a tracer and the integrated sampling method uses perfluorocarbon as a tracer, they are presented according to tracer gas. Other tracers could be used with either method; for example, CO_2 can be used as a tracer and a continuous CO_2 monitor used to measure decay. However, the basic principles of determining the air exchange rate are the same.

Tracer gas dilution: SF_6

Specific instructions for this method can be found in the American Society of Testing Materials (ASTM) Standard Method for Determining Air Leakage Rate by Tracer Dilution (E 741) (Annual Book of ASTM Standards, Volume 4.07, 1983). The basic apparatus for this method includes: tracer gas monitor, cylinder of tracer gas, sample collection containers and pump, syringes, circulating fans, and a stop-watch. Examples of sampling containers are Tedlar^{TM} bags or gas-tight syringes. Meteorological parameters which are recorded include: wind speed and direction, temperature (indoors and outdoors), relative humidity and barometric pressure.

Two types of tracer gas monitors can be used with this technique, depending on the expected concentration of SF_6. Both instruments should be equipped with chart recorders. For SF_6 concentrations in the range of 1–500 ppm, a portable infrared gas analyzer is used. For SF_6 concentrations in the ppb range, a gas chromatograph (GC) with an electron capture detector is used. A field GC is preferable so that the concentration of SF_6 can be immediately verified and optimum sample integrity maintained. However if the laboratory is in reasonable proximity to the site, samples can be transported for analysis. Suitable quality control procedures should be implemented to verify sample integrity in such cases.

The air exchange rate is calculated using the following equation (ASTM, 1983):

$$C = C_0^{-It} \qquad (3.6)$$

where:

C = tracer gas concentration at time t
C_0 = tracer gas concentration at time = 0

I = air exchange rate
t = time

This relationship assumes that the loss rate of the initial concentration of tracer gas is proportional to its concentration. This expression can be shown graphically by plotting the log of concentration versus time, which produces a linear relationship the slope of which is the air exchange rate. The graphical approach allows the researcher to observe the scatter in the data and determine if the assumption of linearity is being met. If the ventilation system recirculates a fraction of the indoor air, then this assumption may not hold. Equation (3.6) can be rearranged to yield the expression:

$$I = (1/t)\ln(C_0/C) \qquad (3.7)$$

The first critical step in the field application of this technique, aside from normal calibration procedures for all equipment, is the calculation of the amount of tracer gas to inject. Because the analytical instruments are highly sensitive to the SF_6 tracer, it is important not to produce concentrations which exceed the calibration range of the instrument. Two equations are presented below which can be used to approximate the initial concentrations for closed and ventilated systems. For closed systems, i.e., buildings with no forced mechanical ventilation systems, the following equation can be used (Nelson, 1972):

$$C = \frac{10^6 V_c}{V_d} \qquad (3.8)$$

where:

C = resultant concentration, ppm
V_c = contaminant volume, ml
V_d = dilution volume, ml

In this equation dilution volume is the volume of the space available for effective mixing of the SF_6. For facilities with forced mechanical ventilation systems, it is necessary to take into account the dilution of the initial concentration due to the supply and exhaust flows. The generalized relationship with initial concentration is shown below (Nelson, 1972):

$$C = C_0 e^{(-V_w/V_c)} \qquad (3.9)$$

where:

C = the resultant concentration
C_0 = the original concentration

Vw = the volume of air removed from the chamber
Vc = chamber volume

Determining the point of release for the tracer depends on the type of structure. In buildings with forced air central heating or air conditioning systems, the tracer can be injected directly into the duct work near the circulating fan. Care should be taken to assure that the duct work is airtight, as leaks can lead to an uneven distribution of the tracer. If there is no central system, the tracer is usually released at one or more points within the structure and circulating fans used to mix the gas. Fans are critical as SF_6 is heavier than air. If it is injected in undiluted form, SF_6 may tend to sink and accumulate in low areas. This stratification will invalidate the relationship between air infiltration and SF_6 decay. With both injection techniques adequate time should be allowed for the tracer to become uniformly distributed. Times can range from a few minutes to one-half hour, depending on the building.

The grab sampling interval is dependent on the expected air exchange rate. In buildings with high rates, long sampling intervals may cause the tracer gas to be diluted too rapidly and there will be too little data to determine adequately the logarithmic linear relationship. Alternatively, if the building has low air exchange rates, short sampling intervals will produce excessive amounts of data and will be inefficient in terms of sampling resources and time. It is important that the maximum concentrations be measured. Thus, frequently the initial samples are taken at fairly short intervals, with longer intervals between the later samples when it is known that the peak initial concentration has passed.

Figure 3.5 is an example of an SF_6 decay curve. It is from a commercial facility which did not recirculate indoor air to this area; it was equipped with a separate exhaust system and conditioned outdoor make-up air. The air exchange rate, estimated from 1.5 minutes to 12.5 minutes, is 4.6 air changes per hour (ACH). The "tail" from 12.5 minutes to 21 minutes may be due to SF_6 escaping to the adjacent hallway and then becoming re-entrained in the office area through the doorways.

When conducting tracer gas measurements, it is important to document various structural parameters and occupant activities which may be occurring during the sampling time as well as the meteorological parameters. Structural parameters include: windows (number, location, type), noticeable leakage paths, wall construction, location of chimneys, vents and other direct indoor-outdoor communication points, and type and capacity of the heating and/or air conditioning systems. Occupant activity such as opening and closing of doors (interior or exterior) or vents will affect the infiltration rate as well as the distribution of the tracer gas within the structure. Operational status of the heating or cooling system should also be recorded. These items do not directly figure into the calculation of the air exchange rate; however, they may be critical in the identification of anomalous data for a particular building and in the comparison of data among buildings.

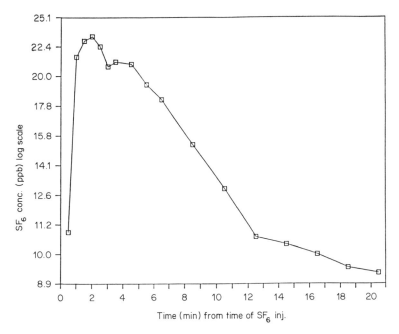

Figure 3.5   Typical tracer decay pattern

Tracer Gas Dilution: Perfluorocarbon

The perfluorocarbon (PF) technique, developed by Dietz and Coté (1982), uses a permeation tube to release this tracer at a known rate for an extended period of time. The method is further described in Dietz *et al.* (1985). The sampling devices which are passive diffusion tubes collect integrated samples over several days. Tubes are analyzed at a laboratory by a gas chromatograph with an electron capture detector. There are two advantages to this technique: it is simple to deploy and it provides a measure of the average infiltration rate over an extended time period. The major disadvantages are that the analytical portion of the method is complex and sophisticated, the cost of PF is high, and few laboratories are equipped to handle these samples on a routine basis.

The required field equipment for this method is the source permeation tube, tracer, and collection tubes. Three different perfluorocarbons have been used: perfluorodimethylcyclohexane, perfluoromethylcyclohexane, and perfluorodimethylcyclobutane. The source is a fluoroelastomer plug impregnated with a known mass of one of these compounds and sealed. The PF diffuses from the plug end at a constant rate and at a concentration which is determined analytically. Dietz and Coté (1982) report emission rates in the range of 3 to 20 nanoliters/minute. The number of sources deployed in a particular building depends on the number of main rooms and an estimation of the probable effective mixing.

Passive samplers are small tubes approximately 3 inches long and 0.25-inch in diameter. The absorbent material, 50 mg of Ambersorb^{TM}, is held in place with glass wool or a stainless steel screen. The number of passive samplers deployed depends on the size and number of rooms in the building. The duration of sampling time depends on several factors including: detection limit of PF on the collection tubes, source emission rate, and field program considerations. Sampling times of four to seven days and up to several weeks have been used successfully with one collection tube. Longer integration times are possible by using sequential collection devices.

The air exchange rate is calculated according to the following equation (Dietz and Coté, 1982):

$$R_v(t) = \frac{V_h}{V(t)}(R_s - \mathrm{d}V(t)/\mathrm{d}t) \qquad (3.10)$$

where:

$R_v(t)$ = infiltration rate, liter/min
V_h = house volume, liter
$V(t)$ = volume of tracer in the house at time t, liter
R_s = tracer source rate, liter/min

To convert this infiltration rate to the more common units of air changes per hour, the infiltration rate ($R_v(t)$) is divided by the house volume (V_h).

3.3.2 Fan Pressurization

The second category of methods considered here is fan pressurization or depressurization. This not a direct measure of infiltration; it characterizes the building leakage rate independent of weather conditions (ASHRAE, 1985). This is done by taking several measurements of the air flowing through the shell of the structure as a function of the pressure difference across the shell. Measurements are made by using a large fan to create an incremental static pressure difference between the interior and the exterior of the building. The air leakage rate is determined by the relationship between the airflow rates and pressure differences. The fan is usually placed in the door, and all direct openings in the building envelope, e.g., windows, doors, vents, and flues, are sealed off. The airflow rate through the fan is determined by measuring the pressure drop across a calibrated orifice plate. The resulting leakage occurs through the cracks in the building envelope, and the effective leakage area can be calculated from the flow profile (Harrje et al., 1981).

There are several advantages of the fan pressurization method: (1) it does not require sophisticated analytical equipment as do the tracer techniques; (2) it allows for a comparison of homes based on their relative leakiness irrespective of the prevailing weather conditions at the time of measurement; and (3) it can be used to measure the effectiveness of retrofit measures. The disadvantage of such an indirect measure of infiltration is that it approximates the actual process through an inherently artificial process, pressurization or de-pressurization.

Canadian and U.S. standard protocols exist for this technique. The Canadian General Standards Board technique is Standard 149.10-M86, "Determination of Airtightness of Buildings by Fan Pressurization Method" (1986). The American Society for Testing and Materials (ASTM) method for this technique, "Determining Air Leakage Rate by Fan Pressurization" (E779–81), is summarized below (ASTM, 1981). The major components of the apparatus for this technique include: a fan or air mover (this may be part of an assembly which will fit into common door openings and is then called a blower door), a manometer or other pressure measuring device with an accuracy of ± 0.01 in. H_2O, airflow or velocity measuring device with an accuracy of $\pm 6\%$ of the average value, and meteorological equipment for recording wind speed and direction, temperature, and relative humidity. The fan or blower should accommodate a wide range of flow rates up to 3000 cfm (5100 m^3/h) and must be calibrated.

General steps to follow in this method include:

• note the physical characteristics of the building,
• close all normal openings (e.g., windows, doors, vents, and flues),
• record meteorological conditions and indoor temperature and relative humidity, and
• install the blower assembly.

The blower should be run at such speeds as to induce pressure differences of 0.05 to 0.3 in. H_2O (12.5 to 75 Pa). Data should be collected at five different pressures and the airflow rate measured at each point. After collecting the data under pressurized conditions, the flow of the fan is reversed, and the measurements repeated for de-pressurized conditions.

ASTM provides guidelines for obtaining data under preferred meteorological conditions. These include: on-site wind speed of less than 5 mph (8 km/h) and indoor–outdoor temperature differences of not more than 20 °F (11 °C). Under these conditions the induced pressure differences between inside and outside the building envelope will be most stable.

After all the data have been collected, the airflow rates are usually converted to a reference condition such as 20 °C and 1 atm. Air leakage is then plotted against the corresponding pressure differences. An example of such a plot for both pressurization and de-pressurization is shown in Figure 3.6. To compare fan pressurization data among buildings, ASHRAE (1985) recommends that the average flow rate at 0.2 in. H_2O (50 Pa) be divided by the volume of the structure to give results in ACH. It should be noted that air infiltration, as ACH, derived in this manner is not equivalent to air infiltration measured via tracer gas and also reported as ACH. ASHRAE (1985) gives the generalization that at similar pressure differences (i.e., those induced by fan pressurization and those induced by normal weather conditions at less than 0.1 in. H_2O), if the leaks are of relatively the same size and uniformly distributed over the structure, the infiltration rate (flow driven by meteorology) will be about one-half the leakage rate (flow driven by pressurization).

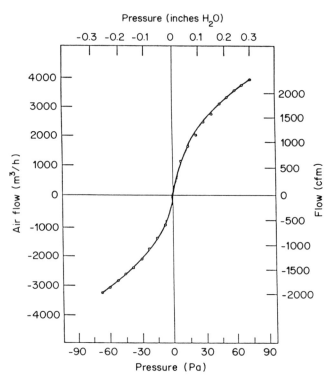

Figure 3.6   Typical flow versus pressure characteristic obtained using fan pressurization (Grimsrud *et al.*, 1983)

3.3.3 Effective Leakage Area

Effective leakage area (ELA) is another indirect method to estimate air infiltration. ELA can be interpreted physically as an approximation of the total area of physical openings in the building envelope through which infiltration occurs. The empirical model used to estimate air exchange is based on pressure differences and was developed by several researchers at Lawrence Berkeley Laboratory (Grimsrud *et al.*, 1983; Sherman, 1987). The model predicts air exchange on the basis of the following parameters:

- leakage area of the structure and its distribution,
- geometry of the structure,
- inside–outside temperature difference,
- wind speed,
- terrain class of the location, and
- shielding class around the structure.

The LBL model assumes: the structure is a single-cell, and the leaks in the envelope can be treated as simple orifices. This latter assumption means that the flow due to each driving force, wind, temperature difference, or mechanical ventilation can be calculated and then combined (Sherman, 1987). Conceptually, the model uses an effective leakage area value and predicts the volumetric flow through this area from wind pressure and then from temperature difference or stack effect. These two values are then combined in quadrature. The model is presented in the following equations (Sherman and Modera, 1986):

Stack-induced infiltration, Q_s:

$$Q_s = L_o f_s \left(g H_s \left| \frac{\Delta T}{T} \right| \right)^{0.5} \tag{3.11}$$

where:

L_o = leakage area, m^2
f_s = stack factor
g = acceleration of gravity, 9.8 m/s^2
H_s = "stack height" of building (distance between lowest and highest leak)
ΔT = inside–outside temperature difference, K
T = reference indoor temperature, 295K

Wind induced infiltration, Q_w:

$$Q_w = L_o f_w v \tag{3.12}$$

where:

f_w = wind factor
v = wind speed, m/s

Total flow, Q:

$$Q = (Q_s^2 + Q_w^2)^{0.5} \tag{3.13}$$

For buildings with a mechanical ventilation system, the following expression applies:

$$Q = Q_{bal} + (Q^2_{unbal} + Q^2_s + Q^2_w)^{0.5} \qquad (3.14)$$

where:

Q_{bal} = minimum of either supply or exhaust flow
Q_{unbal} = maximum of either supply or exhaust $- Q_{bal}$

The stack factor is a function of the neutral pressure level and the fraction of total leakage in the floor and ceiling. The wind factor is a function of the terrain class of the location, wind speed on-site (or corrected to on-site, if local weather station data are used), and the fraction of leakage in the floor and ceiling. These values can be approximated as described below. The interested reader is referred to Sherman and Modera (1986) for a complete discussion of the derivation of the expressions for f_s and f_w.

Leakage area is the primary parameter which must be quantified to use this model. Grimsrud et al. (1983) present a method for calculating pressure difference across the building envelope based on the leakage area of the various openings. The method involves measuring the dimensions of each opening and converting this value to a leakage area equivalent value. The conversion is done via numerous ancillary tables (provided in Grimsrud et al., 1983). The interested reader should consult this reference and Lawrence Berkeley Laboratory for more detailed information on this technique.

Sherman (1987) presented a simplified approach to estimating air infiltration from fan pressurization data which is based on the model presented by Grimsrud et al. (1983). Sherman (1987) notes two caveats which apply to this approach: first, it is best suited for large data sets which may not have the accuracy of intensive, smaller scale studies, and second, it produces a value which is representative of a yearly average air infiltration rate as opposed to a measurement at one point in time. These restrictions do not necessarily apply to the model presented by Grimsrud et al. (1983). The version of the LBL model as presented by Sherman and Modera (1986) can be used to calculate air exchange rates over time periods of less than one hour.

Sherman's approach is presented in two steps: estimating the surface area of the leakage paths in a building (effective leakage area) and multiplying that value by an estimate of the velocity of air which would pass through the opening. The effective leakage area (ELA) is calculated from fan pressurization data. It is defined by the following equation:

$$ELA = \frac{Q_4}{(2\Delta P / \rho)^{0.5}} \qquad (3.15)$$

where:

> ELA = effective leakage area, m²
> Q_4 = airflow at 4 Pa (m³/sec)
> ΔP = the pressure drop causing this flow, i.e., 4 Pa
> ρ = density of air, 1.2 kg/m³

The second step is the calculation of the specific infiltration, s. This linear velocity term is a function of both wind speed and indoor–outdoor temperature difference. Although building dependent parameters would influence this value, they are not taken into account in this simplified approach. The equation for calculating the specific infiltration is as follows:

$$s = (f_w{}^2 v^2 + f_s{}^2 |\Delta T|)^{0.5} \tag{3.16}$$

where:

> f_w = the wind parameter
> v = the wind speed, m/s
> f_s = the stack parameter, m/s*K$^{0.5}$
> ΔT = absolute value of the indoor–outdoor temperature difference, °K

Sherman (1987) states that f_w and f_s depend on the leakage distribution and siting of the particular structure being modeled. He has compiled data on a substantial number of homes and determined that typical single-family houses have values of:

$$f_w = 0.13$$

$$f_s = 0.12$$

Temperature difference values and wind speed values can be estimated from local airport data or other meteorological recording stations. Depending on the averaging period of interest for the structure, appropriate average wind speed and temperature difference values should be selected. If the meteorological data are taken from remote distances and/or there are significant variations in terrain between the station and the building of interest and/or significant shielding (trees or other structures) around the building, a correction to wind speed values should be made. Grimsrud et al. (1983) present one way of making this correction.

The average air infiltration, Q, for a structure is calculated as follows:

$$Q = \text{ELA} \times s \tag{3.17}$$

where:

Q = infiltration, m^3/s
ELA = effective leakage area, m^2
s = specific infiltration, m/s

To convert to units of air changes per hour (ACH), Q is converted to cubic meters per hour and divided by the volume of the structure (in meters).

Limited studies have been conducted comparing the two versions of the LBL model described above to tracer gas measurements of air exchange rates. Coon (1984) reported differences of 35–47% between tracer gas data and the model of Grimsrud *et al.* (1983). The full LBL model has been evaluated by Liddament and Allen (1983) and Sherman and Modera (1986). Liddament and Allen reported that the model predicted air exchanges to within ±25% of tracer gas measurements. Sherman and Modera (1986) reported an accuracy of ±20% for short-term measurements, but accuracy increases to ±7% when the averaging period is one week.

3.3.4 Summary of Measurement Techniques

Three of the more common techniques for quantifying air infiltration have been presented: tracer gas dilution, fan pressurization, and effective leakage area. Tracer gas dilution is a direct measurement of air exchange, whereas the other methods are indirect. If employed properly, tracer gas dilution is an effective technique for measuring the air exchange under a given set of meteorological conditions. It is not certain how representative a one-time measurement is of the average infiltration rate of a structure. Fan pressurization data are independent of weather conditions as it measures airflow through the building envelope under artificial conditions. Although the test is independent of actual meteorological conditions, it is not independent of seasonal effects; e.g., low versus high humidity can affect a fan pressurization value by as much as ±40–50%. Effective leakage area is an empirical approach to calculating air exchange from fan pressurization data; the accuracy of this approach has yet to be reported. The advantage of this last technique is that it can be used to generate seasonal or yearly values by using the appropriate meteorological data.

A word of caution is required in interpreting air exchange rate data. Because tracer gas dilution and fan pressurization measure different physical properties related to air infiltration, data from these two methods are not directly comparable. Even though data from fan pressurization tests can be used to generate an air exchange rate value (as ACH), this value is not equivalent to an air exchange rate value (as ACH) measured by tracer gas dilution. Thus when reviewing the literature or reporting results of air exchange measurements, the data from each method should not be combined without adequate justification.

3.4 Examples of Air Exchange Rate Measurement Data

Numerous studies have been reported which collected air exchange rate data as part of the research effort. It is beyond the scope of this work to present a thorough review of such data; however, a few examples of air exchange rate measurements are presented.

One of the more extensive reviews of air infiltration measurements is a Canadian survey published by the Ontario Ministry of Municipal Affairs and Housing (Coon, 1984). For measurements using fan pressurization, one study reported results from 1976 homes in Saskatoon, Canada. The fan pressurization data were normalized to account for differences in the size of the homes. This was done by dividing the effective leakage area by the above ground surface area of the building envelope; the units are reported as cm^2/m^2 and called "specific leakage area." The Canadian homes were grouped according to age and the data are shown in Table 3.1.

Table 3.1

Age	Specific leakage area (cm^2/m^2)
Pre–1945	4.5–10.0
1945–1960	2.5–4.5
1960–1980	1.5–2.5
"Low energy"	1.5

Source: D. Coon, "Indoor Air Quality in Tight Houses: a Literature Review," Housing Conservation Unit, Ontario Ministry of Municipal Affairs and Housing, Ontario, Canada, 1984; "low energy" not defined.

One conclusion drawn from this study was that age of the structure had an effect on the specific leakage area.

In another study of 35 new "energy efficient" homes in the Ottawa area, air infiltration was measured by tracer gas dilution (SF_6 and PF). Measurements were conducted in the winter, and the author notes that these homes are tighter than most newly constructed homes. These results are summarized in Table 3.2.

Table 3.2

Heating system	Number	Air changes/h (ACH)	
		Mean	Range
Gas	31	0.31	0.18–0.55
Electric	4	0.21	0.09–0.31

Source: D. Coon, "Indoor Air Quality in Tight Houses: a Literature Review," Housing Conservation Unit, Ontario Ministry of Municipal Affairs and Housing, Ontario, Canada, 1984.

PF was used to measured air infiltration over a two-week period in a subset of seven gas heated homes. Values ranged from 0.09 to 0.31 ACH, with a mean of 0.18 ACH.

Grimsrud *et al*. (1983) compiled data from numerous studies conducted in the United Sates and Canada. Of the 312 homes reported, fan pressurization data were available for 287 homes. The Lawrence Berkeley Laboratory model (cited above) was used to calculate average air infiltration. In addition, the specific leakage area was also reported. Data were all taken during the timeframe of November to March. The authors provide an extensive listing of all the data derived from this survey of the literature. However, the only summary values presented are for air exchange rate, a mean of 0.63 ACH and a median of 0.50 ACH.

Tracer gas measurements (SF_6) were reported by Lamb *et al*. (1985) for 10 homes in the state of Washington as part of a weatherization program. The data were collected over different seasons, with air exchange rates varying from 0.3 to 1.0 ACH for "typical meteorological conditions" which are not defined. An interesting observation from this study was that in one home the extended use of doors caused the infiltration rate to increase three-fold.

3.5 Conclusions

Air infiltration is a dynamic property; the rate at which the air inside a structure is exchanged with outdoor air is strongly dependent upon construction, occupant living patterns, indoor–outdoor temperature differences, wind speed, and wind direction. With respect to meteorological conditions, larger temperature differences, as well as faster wind speeds, increase the air infiltration rate. Wind direction effects are usually less important and are dependent on the orientation of the leakage paths relative to the prevailing wind direction and various shielding factors surrounding the structure.

Different techniques have been developed to measure air infiltration, either directly or indirectly. Empirical expressions have been derived to relate leakage area measurements to infiltration by taking into account meteorological conditions. These models involve the use of empirical constants. There has been limited testing of these models against actual measurements of air exchange rate. Further research is needed comparing air exchange rate values determined by different techniques and under different meteorological conditions.

References

ASHRAE (1981). *Fundamentals Handbook*, American Society of Heating, Refrigerating and Air Conditioning Engineers, Inc.
ASHRAE (1985). *Fundamentals Handbook*, American Society of Heating, Refrigerating and Air Conditioning Engineers, Inc.

ASHRAE (1989). *Fundamentals Handbook*, American Society of Heating, Refrigerating and Air Conditioning Engineers, Inc.

ASTM (1981). "Standard Test Method for Determining Air Leakage Rate by Fan Pressurization," Designation E 779–81, in *Annual Book of ASTM Standards*, American Society for Testing and Materials, Vol. 4.07.

ASTM (1983). "Standard Test Method for Determining Air Leakage Rate by Tracer Dilution," Designation E 741–83, in *Annual Book of ASTM Standards*, American Society for Testing and Materials, Vol. 4.07.

Alevantis, L.E., and Girman, J.R. (1989). "Occupant-Controlled Residential Ventilation," *IAQ-89 The Human Equation: Health and Comfort*, American Society of Heating, Refrigerating and Air Conditioning Engineers, Atlanta, GA.

Blomsterberg, A.K., and Harrje, D.T. (1979). "Evaluating Air Infiltration Energy Losses," *ASHRAE Journal* 21:25–32.

Brundrett, G.W. (1977). "Ventilation: A Behavioral Approach," *International Journal of Energy Research* 1:289–298.

Canadian General Standards Board (1986). "Determination of Airtightness of Buildings by the Fan Depressurization Method," Standard no.149.10-M86, Canadian General Standards Board, Ottawa.

Charlesworth, P.S. (1988). "Air Exchange Rate and Airtightness Measurement Techniques—An Applications Guide," International Energy Agency, Air Infiltration and Ventilation Centre, Coventry, Great Britain.

Coon, D. (1984). "Indoor Air Quality in Tight Houses: A Literature Review," Prepared for Housing Conservation Unit, Ontario Ministry of Municipal Affairs and Housing, Ontario, Canada.

Dietz, R.N., and Coté, E.A. (1982). "Air Infiltration in a Home Using a Convenient Perfluorocarbon Tracer Technique," Dept. of Energy, Brookhaven National Laboratory, Upton, New York.

Dietz, R.N., Goodrich, R.W., Coté, E.A., and Wieser, R.F. (1985). "Detailed Description and Performance of a Passive Perfluorocarbon Tracer System for Building Ventilation and Air Exchange Measurements," Dept. of Energy, Brookhaven National Laboratory, Upton, New York.

Grimsrud, D.T., Sherman, M.H., and Sonderegger, R.C. (1983). "Calculating Infiltration: Implications for a Construction Quality Standard," U.S. Dept. of Energy, Lawrence Berkeley Laboratory Report LBL-9416, Berkeley, CA.

Harrje, D.T., Persily, A.K., and Linteris, G.T. (1981). "Instruments and Techniques in Home Energy Analysis," Presented at IEA Energy Conservation in Buildings and Community Systems," Elsinor, Denmark.

Lamb, B., Westberg, H., Bryant, P., Dean, J., and Mullins, S. (1985). "Air Infiltration Rates in Pre- and Post-Weatherized Houses," *J. of the Air Pollution Control Assoc.* 35:545–555.

Liddament, M. (1986). "Air Infiltration Calculation Techniques—An Applications Guide," International Energy Agency, Air Infiltration and Ventilation Centre, Coventry, Great Britain.

Liddament, M., and Allen, C. (1983). "The Validation and Comparison of Mathematical Models of Air Infiltration," International Energy Agency. Technical Note AIC 11, Air Infiltration Centre, Bracknell, Berkshire, Great Britain.

Liddament, M., and Thompson, C. (1983). "Techniques and Instrumentation for the Measurement of Air Infiltration in Buildings," International Energy Agency, Air Infiltration Centre, Technical Note 10, Bracknell, Berkshire, Great Britain.

Luck, J.R., and Nelson, L.W. (1977). The Variation of Infiltration Rate with Relative Humidity in a Frame Building," *ASHRAE Transactions*, Vol. 83, Part I, 718–729.

Lyberg, M.D. (1981). "Energy Losses Due to Airing by Occupants," Swedish Institute for Building Research, Gavle, Sweden.

Nelson, C.O. (1972). *Controlled Test Atmospheres, Principles, and Techniques*, Ann Arbor Science Publishers, Inc., Ann Arbor, Michigan.

Sherman, M.H. (1987). "Estimation of Infiltration from Leakage and Climate Indicators," *Energy and Buildings* **10**:81–86.

Sherman, M.H., and Modera, M.P. (1986). "Comparison of Measured and Predicted Infiltration Using the LBL Infiltration Model," in *Measured Air Leakage of Buildings*, ASTM STP 904, H.R. Trachsel and P.L. Lagus (eds), American Society for Testing and Materials, Philadelphia, PA.

Stricker, S. (1975). "Measurement of Airtightness of Houses," *ASHRAE Transactions*, Vol. 81, 148–167.

Tucker, W.G. (1990). Personal communication to J.E. Yocom, 17 January, 1990.

Yarmac, R.F., McCarthy, S.M., and Yocom, J.E. (1987). "Final Report on Air Unvented Gas Space Heater Study," prepared for the Consumer Product Safety Commission, Washington, D.C.

Yocom, J.E., Murphy, B.L., and Bicknell, B.R. (1986). "Final Report on Environment—Indoor Migration Factors," EPA Contract No. 68–03–3233, U.S. Environmental Protection Agency, Athens, GA.

CHAPTER 4

Methods for Measuring Indoor Pollutants

In the United States there are well established methods for the measurement of air quality in both the outdoor ambient atmosphere and in the industrial occupational atmosphere. For example, the U.S. EPA has established Reference Methods for all of the air pollutants for which outdoor ambient air quality standards have been developed (U.S. Code of Federal Regulations, Part 50.1(e), 1989). Recently, the U.S. EPA has initiated an effort to produce a compendium of Indoor Air Quality Methods which will continue over the next few years. As of 1988, two methods have been published, one for volatile organic compounds (VOCs) and one for nicotine (Winberry et al., 1988).

In the case of air quality in the occupational setting, the U.S. National Institute of Occupational Safety and Health (NIOSH) has developed standard methods for measurement of airborne contaminants in the industrial workplace (NIOSH, 1984). These methods are used to determine if concentrations of air contaminants in occupational settings comply with Permissible Exposure Limits (PEL) adopted by the U.S. Occupational Safety and Health Administration (OSHA).

In addition to these governmentally sanctioned methods, Committee D-22 of the American Society of Testing Materials (ASTM) has prepared standard methods for measuring many of the outdoor ambient and occupational pollutants mentioned above. A new subcommittee (D-22.05) of this organization was formed in 1986 to develop standard methods for measurement of indoor air quality. This group is adapting both outdoor ambient and indoor occupational methods for application to the non-occupational and non-industrial occupational (e.g., office buildings) settings. Anyone seriously interested in indoor air quality monitoring methods should keep abreast of the activities of this group.

In general, the concentrations of pollutants of concern in indoor air quality are in the range found in the outdoor atmosphere, but in some instances (e.g., in homes with unvented or poorly vented combustion sources where extremely high concentrations of CO and NO_2 may be found) indoor concentrations may approach occupational levels. Nevertheless, measurement methods designed for monitoring outdoor ambient concentrations have greater applicability to the indoor environment than methods designed for measuring industrial occupational concentrations. In fact, many of the major indoor air quality monitoring studies have relied heavily on outdoor monitoring methods modified as necessary for indoor monitoring. These outdoor measurement methods are generally of the integrated grab sample or continuous monitoring and recording type.

Most of the NIOSH occupational air sampling methods (NIOSH, 1984) are based on integrated grab sampling. Many of these methods use bubblers to contact sample air with absorbent solutions which are subsequently analyzed for the absorbed pollutants or their reaction products. Methods for sampling contaminants in particulate form are usually based on filtration followed by chemical analysis. Other methods, such as those used for organic vapors, rely upon adsorption on solid sorbents followed by analysis or direct measurement by physical measurement (e.g., CO by infrared spectrophotometry). While these methods were originally developed for measuring the relatively high concentrations encountered in industrial occupational exposure, they can be extremely useful to the indoor air quality researcher in situations where established indoor or outdoor air quality methods are not yet available. NIOSH has published methods for hundreds of air contaminants which are all theoretically available for indoor air quality work, but in applying them one must consider a number of important factors, including:

- Applicability of the method to required concentration ranges.
- Feasibility of increasing the sensitivity of the method without degrading precision and accuracy (e.g., increasing sampling times).
- Specificity of method and freedom from interferences.
- Possible generation of contaminants and noise that could affect the indoor living or working environment.

The American Conference of Governmental Industrial Hygienists (ACGIH) has been active in the development of methods applicable principally to occupational exposures. At a symposium held in 1987 under the sponsorship of ACGIH, advances in air sampling technology were discussed. Many of the methods presented are applicable to non-occupational indoor exposures (ACGIH, 1989a).

When considering indoor air quality monitoring for estimating human exposure to pollutants, one must be aware of the distinction between fixed point and portable or personal monitoring. Fixed point monitoring is intended to depict air quality at fixed points either outdoors or indoors. Outdoor ambient monitoring is usually carried out in this way using continuous or integrating (passive or active) monitors. Portable or personal monitors are used to depict the integrated exposure of an individual as he or she passes through various microenvironments. (A "micro-environment" is a location characterized by relatively homogeneous pollutant concentration (e.g., home, office, auto, or subway) which a person occupies for a finite period of time in the course of his or her normal daily activities.)

Fixed point sampling can be carried out in only one of the individual micro-environments. If one wishes to determine total exposure from time spent in a number of microenvironments and fixed point sampling is carried out in each of the microenvironments, total exposure can only be estimated indirectly from data collected in personal activity logs. On the other hand, data from personal monitors, where the individual carrying the monitor passes through a series of microenvironments in the course of a day or some other sampling interval, produces total exposure to the pollutant being monitored, but it is often impossible to separate out exposures in individual microenvironments. This is especially true when integrating samplers are used.

Another type of measurement approach is personal monitoring with compact, lightweight systems that can be attached to a human subject to obtain a picture of total exposure to a pollutant. Until fairly recently, these sampling systems have been limited to providing only integrated samples over a significant sampling period (e.g., 8, 12, or 24 hours), but now with electronic miniaturization and use of chip-based data loggers, it is possible to obtain reasonably continuous records of pollutant exposures for several pollutants. Figure 4.1 is a diagram showing the status of development of personal monitors in the U.S. (U.S. EPA, 1988a). Earlier, Wallace and Ott (1982) conducted a state-of-the-art survey of

Pollutants		Monitor needed	Monitor under development	Prototype monitor development	Tested and evaluated	Used in pilot studies	Used in large field studies	Ready for routine use
CO	D							→
	I		→					
NO₂	D				→			
	I						→	
Inhalable particles	D			→				
	I						→	
Formaldehyde	D	→						
	I						→	
VOCs	D	→						
	I						→	
Pesticides	D	N/A						
	I						→	
Radon	D							→
	I							→
PAH	D	N/A						
	I		→					
Biological aerosols	D	N/A						
	I							→
House dust	D	N/A						
	I		→					

Figure 4.1 Status of personal monitor development. Reproduced from U.S. EPA (1988a). D, direct readout; I, integrating collection of samples; N/A, not applicable

personal monitors. More recently, Lewis (1989) reviewed work at EPA to develop and evaluate indoor air quality monitoring devices.

The purpose of this chapter is to familiarize the reader with available methods for specific pollutants and their operating principles and provide an overview of measured indoor concentrations as reported in the literature. The field of indoor air quality monitoring is developing rapidly. The authors do not present detailed descriptions of commercially available equipment because new instruments are constantly being brought to the market. Rather, we will discuss available measurement methods and concepts and trends in new developments as of the time of the publication of this book. An excellent review of commercially available equipment for indoor air quality and personal exposure monitoring in the approximate time period 1985–6 is contained in the book by Nagda et al. (1987).

The organization of this chapter is by pollutant or pollutant category, starting with those which are generated outdoors or both outdoors and indoors and ending with those that are primarily of indoor origin. Each section on a given pollutant contains a review of the sources of the pollutant, available measurement methods, and a review of the results of the most significant indoor air quality measurement programs. Summaries of the data from representative major studies, organized by pollutant, are presented in the appendices. The reader is directed to the literature cited to learn the details of how a given monitoring technique was employed. The references cited in the appendices are included in the reference list at the end of this chapter.

4.1 Carbon Monoxide (CO)

4.1.1 Sources and Characteristics of CO

As was shown in Table 1.3, CO is generated both outdoors and indoors and is the product of incomplete combustion. The primary outdoor source is motor vehicle exhaust. Indoor sources include unvented or improperly vented combustion devices (e.g., gas stoves and heaters and kerosene heaters). Other indoor sources include wood stoves and cigarette smoking. The importance of CO as an air pollutant is related to its ability to interfere with the oxygen transport capacity of the blood. The affinity of hemoglobin for CO is over 200 times that of oxygen. Therefore it takes only a few parts per million of CO to be of health concern.

CO is an unreactive gas and readily penetrates from outdoors without undergoing significant depletion by physical and chemical processes other than by dilution through air exchange. Once it is present in the indoor air, whether from outdoor or indoor sources, it can be removed only by exchange with fresh, CO-free air. Its stability often makes it useful as an indoor tracer for air exchange determinations. However, one must be careful in any experiment in which indoor air is "spiked" with CO to keep the indoor concentrations at acceptable exposure levels such as the outdoor air quality standards.

4.1.2 Available Measurement Methods for CO

The most commonly used method for measuring CO at fixed points indoors and outdoors is by means of a non-dispersive infrared (NDIR) analyzer. The NDIR principle is the basis for one of the EPA Reference Methods. Infrared energy from a source is passed through a cell containing the gas to be analyzed, and the quantitative absorption of energy by CO in the sample cell is measured by a suitable detector. The photometer is sensitized to CO by employing CO gas in either the detector or in a filter cell in the optical path, thereby limiting the measured absorption to one or more of the characteristic wavelengths at which CO strongly absorbs. Optical filters or other means may also be used to limit sensitivity of the photometer to a narrow band of interest.

The U.S. EPA has recognized the gas filter correlation (GFC) method as equivalent to NDIR for measurement of CO. GFC is also a method based on infrared radiation, but in this case the radiation passes through a spinning filter wheel that contains CO and nitrogen reference cells. The infrared radiation then passes through a detector cell containing the air being sampled. The differences in signal between the nitrogen cell and the CO cell are proportional to the CO concentrations in the sample.

At present there are six commercially available instruments based on NDIR and two based on GFC. Since the list of reference methods is constantly being revised, no attempt is made to provide the list current as of the writing of this book. Rather, the reader is directed to the most current issue of the U.S. Code of Federal Regulations and to the U.S. EPA Atmospheric Research and Exposure Assessment Laboratory in Research Triangle Park, North Carolina to obtain the most up-to-date information.

CO monitors using NDIR have proven reliability and sensitivity and can provide a continuous record of CO concentrations, but they are much too bulky to serve as personal exposure monitors. In the early 1980s several new miniaturized CO monitoring instruments became available as described by Wallace and Ott (1982). These devices could be hand-held, belt mounted, or carried over the shoulder like a camera or purse. At the same time, several passive CO detectors were developed. These monitoring devices made it possible to conduct human exposure studies for CO. The most successful operating principle for these devices is electrochemical or catalytic oxidation of CO. A number of U.S. companies are currently manufacturing instruments based on this principle.

Ott *et al*. (1986) describe the work done by the U.S. EPA to add microprocessor data loggers to an existing General Electric (GE) electrochemical CO monitor. The GE instrument is based on the oxidation of CO to CO_2 in a proprietary solid polymer electrolyte. The electric signal produced is proportional to the concentration of CO in the sample air. CO concentration is read directly from a liquid crystal display. A chemical filter removes interferences such as NO_2.

Two CO Exposure Dosimeters (COED) were developed: COED-I and COED-

II (Ott *et al*., 1986). The data logger on COED-I permitted the recording of CO concentrations over one-hour periods or during each period representing a different activity. However, those carrying and operating the devices had to record activities in a diary. Figure 4.2 is a diagram of the COED-I system. The COED-II was fitted with a more sophisticated data logger, interfaced with a programmable calculator making it possible to program in a variety of activities such as "at work," "commuting," etc. Both versions were successfully used in human exposure studies conducted in Denver, Colorado, and Washington DC (Akland *et al*., 1984).

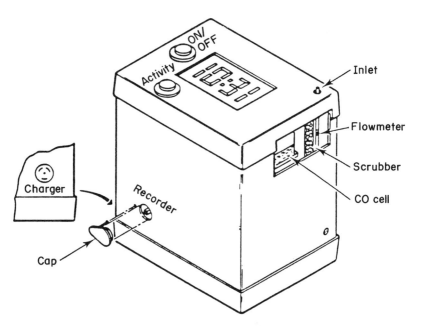

Figure 4.2 Schematic drawing of EPA COED-I CO personal exposure monitor. (Reproduced from Ott *et al*. (1986) with permission of the Air and Waste Management Association)

4.1.3 Indoor Concentrations of CO

Many studies have been carried out to measure CO in several types of indoor environments using a variety of monitoring techniques. CO has been measured in homes, office and public buildings, and automobiles. Monitoring techniques have included fixed point monitors and small portable samplers used at fixed points and as personal exposure monitors (PEMs). The data based on CO studies is fairly large, and continues to grow. Appendix A summarizes the results of major studies which have been published at the time of this writing. See U.S. EPA (1988a) for a more in-depth review of these studies. The highlights of these studies are summarized below:

- The greatest number of studies have focused on the most ubiquitous source of CO, motor vehicle exhaust. Exposure inside vehicles has been studied most frequently; however, exposure in buildings adjacent to areas of motor vehicle exhaust has also been studied. The studies inside vehicles most frequently used personal exposure monitors based on electrochemical principles. Concentrations within the vehicles were dependent on both the density and speed of surrounding traffic and to a lesser extent on the type of vehicle being studied. Reported concentrations were generally in the 10 to 30 ppm range, with some peaks as high as 45 to 55 ppm.
- CO concentrations in buildings have been measured with both fixed monitors (e.g., NDIR) and PEMs. Proximity to traffic and season influence the indoor CO levels. Indoor concentrations are less than outdoor concentrations, with variations attributed to source and ventilation variables. Measured levels in commercial office buildings range from 1 to 10 ppm. Parking garages and other "indoor" vehicle areas can be two to three times higher.
- The third type of study focused on CO levels related to unvented indoor combustion appliances. Studies have been conducted in field settings as well as in controlled laboratory test chambers. In homes, concentrations are highest in the room with the combustion appliance. Depending on the type of home and appliance, concentrations in the range of 6 to 10 ppm are common.
- Research on space heating appliances has shown that emission rate is dependent on the various design (e.g., type of burner head in kerosene stoves) and operational procedures (e.g., wood burn rate in stoves).

The trend in indoor research on CO is to focus on monitoring exposures of sensitive subpopulations and relating these exposures to possible health effects. Environmental tobacco smoke as a source of CO exposure is covered in Section 4.10.

4.2 Nitrogen Dioxide (NO_2)

4.2.1 Sources and Characteristics of NO_2

NO_2 is the most important of a number of nitrogen oxides (NO_x) that may be present as a common contaminant in air. Nitrogen oxides, principally NO_2 and NO, are produced in combustion processes where air is used as the oxidizer. Nitrogen oxides are emitted into the outdoor atmosphere from natural sources (e.g., lightning and forest fires) and from anthropogenic sources (e.g., fuel-fired power and heating plants and motor vehicles). Although NO tends to be produced in greater amount than NO_2 in most combustion processes, NO_2 is of greater importance since it is a respiratory irritant. Both NO_2 and NO are involved in photochemical reactions with reactive organic gases and vapors to produce ozone (O_3) and a complex series of organic gases and aerosols commonly called "smog." Aside from its role as a respiratory irritant, NO_2 indoors is capable of damaging materials; for example, it causes fading of certain types of fabric dyes and has been involved in damage to sensitive electronic equipment (Yocom *et al.*, 1986).

Nitrogen oxides are produced indoors from unvented or poorly vented combustion devices and by tobacco smoking. Small but variable amounts may also be produced by electrical equipment and certain types of office equipment (e.g., photocopiers). Indoor concentrations of NO_2 in homes where unvented combustion devices are used invariably exceed outdoor concentrations and may reach levels of health concern. NO_2 and NO produced outdoors penetrate the indoor environment, but since NO_2 is far more reactive than NO, it decays rather rapidly in the indoor environment. Yarmac *et al.* (1987) compared NO_2 decay rate measured by a number of workers and found that it varies over a range of 0.83 to 1.34 per hour. Billick and Nagda (1987) and Spicer *et al.* (1987) measured NO_2 decay rates in environmental test chambers containing various materials. Decay rates of NO_2 in the presence of various household materials varied over the range of 0 to 8 per hour.

4.2.2 Available Measurement Methods for NO_2

The U.S. EPA Reference Method for NO_2 is gas-phase chemiluminescence. In this method the emission of photons from the gas-phase reaction of NO with ozone is measured. Total oxides of nitrogen (the sum of NO and NO_2 or NO_x) are measured first by reducing all NO_x to NO. NO_2 is then determined by subtracting the quantity of NO from NO_x. At present there are eight commercially available instruments that have received EPA approval. Instruments based on chemiluminescence have been widely used in indoor air quality studies and are particularly useful where relationships between NO and NO_2 and NO_2 decay rates are desired. Small amounts of ozone are released by instruments using this measurement principle, and their use indoors may require outdoor venting.

As in the case of CO, the EPA Reference and equivalent methods are too bulky for portable indoor air or total exposure sampling. In recent years, more and more indoor NO_2 studies have used the Palmes tube (Palmes et al., 1976). This is an integrative passive sampling device based on the principle of molecular diffusion. The sampler consists of three stainless steel screens coated with trimethanolamine (TEA) positioned at one end of a plastic tube, capped when not in use. Figure 4.3 is a diagram of a Palmes tube. When the cap is removed to place the sampler in operation, the NO_2 in the air diffuses towards the coated screens at a rate dependent upon Fick's First Law of Diffusion. The amount of NO_2 measured is a function of its concentration in the air and the length of time the tube is left open. The NO_2 collected is measured in the laboratory by spectrophotometry. Since diffusion is temperature-dependent, these tubes have a rather limited temperature

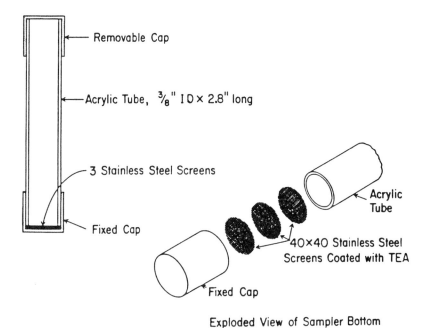

Exploded View of Sampler Bottom

Figure 4.3   Schematic drawing of passive NO₂ sampler. (Reprinted with permission by
American Industrial Hygiene Association Juuornal, Vol. 37, 572 (1976))

range: 15 to 30 °C. Thus outdoor or total exposure sampling in cold weather is not
recommended. Moschandreas *et al.* (1989) evaluated the performance of Palmes
tubes under various climatic conditions and found that low temperatures and high
relative humidities adversely affect the reliability of this device.

Since the Palmes tube was originally developed for industrial hygiene sampling
over 8-hour periods of much higher concentrations than normally encountered in
indoor environments, the device must be exposed for a minimum of 5 to 7 days
for most indoor air quality studies. The lower detectable limit is reported to be
1.5 ppb or three times the blank value for a one-week exposure (Nagda *et al.*,
1987). Nevertheless, the original Palmes tube and variations thereof have found
wide use because of their simplicity and the straightforward chemical analysis.
Several companies now sell the tubes along with the analytical service at reasonable
prices. Recent work at the U.S. EPA has produced an NO₂ passive monitor that
requires only a few hours of exposure to obtain measurable concentrations in non-
occupational indoor environments (Lewis, 1989).

Nagda *et al.* (1987) list a number of commercially available portable
NO₂ monitoring devices including automated wet chemical, chemiluminescent,
electrochemical, and other types of molecular diffusion passive monitors. The U.S.

EPA (1988a) is actively pursuing research towards the development of improved indoor and personal air monitors for NO_2. These include the following:

- Real time monitors based on the chemiluminescent reaction between NO_2 and luminol (Wendel et al., 1983) and the commercial development of a small lightweight monitor (Schiff et al., 1986).
- Development of an electrochemical sensor system with improved sensitivity. Thus far interference from sulfur compounds is a problem.
- Improvements in the sensitivity of the Palmes tube (Miller, 1987).
- Development of an integrative passive monitor based on a sorbent cartridge originally used by Fine (1975) for collection and analysis of nitrosamines.

4.2.3 Indoor Concentrations of NO_2

Since the early 1970s a number of studies have been carried out to measure indoor concentrations and indoor–outdoor relationships of NO_2. Since many of these studies have used chemiluminescent monitors that have the capability of measuring both NO_2 and NO, a significant amount of NO data is also available. More recently, many studies have used the passive monitor, specific for NO_2 alone, developed by Palmes et al. (1976) which can be used either as a fixed or personal monitor. A more complete review of the recent studies on NO_2 is presented in Appendix B.

Research on NO_2 can be broadly grouped into two categories based on source type: gas stoves and other combustion sources. In general, indoor concentrations are less than outdoor concentrations in microenvironments without combustion sources. Most of these studies used passive samplers; only two studies used continuous monitors. The micro-environments studied included homes (kitchen, living-room, bedroom), offices, and schools. Personal monitoring was also frequently conducted. The studies showed that generally when indoor sources were present, the relative rankings of concentrations were: indoor greater than personal greater than outdoor.

The findings from these studies are highlighted below:

- The studies focusing on gas stoves usually monitored in the kitchen and another living area (e.g., living room). Reported concentrations in the kitchen ranged from 9–598 $\mu g/m^3$. Peak concentrations frequently exceeded the annual NAAQS of 100 $\mu g/m^3$. The highest levels were associated with gas stoves in low income apartments. Concentrations in the other areas of the same dwellings were lower, ranging from 20–84 $\mu g/m^3$.
- Concentrations appear to be dependent on duration of stove use, ventilation patterns in the kitchen and rest of the dwelling. One study did not find substantially different concentrations between natural gas and liquified propane gas.
- Other indoor combustion sources which have been studied include: kerosene heaters, unvented gas space heaters, and cigarettes. Concentrations in rooms with space heaters ranged from 20–295 $\mu g/m^3$, depending on the number and type of

heater. One study on unvented gas space heaters reported NO_2 levels of 94–231 $\mu g/m^3$. Cigarettes were found to make a minimal, but detectable contribution of NO_2.

4.3 Sulfur Dioxide (SO_2)

4.3.1 Sources and Characteristics of SO_2

Table 1.3 states that SO_2 is generated primarily outdoors. Its principal source is from the burning of sulfur-containing fuels. Where there are no indoor sources, indoor–outdoor ratios from SO_2 are generally 0.5 or less due to the reactivity of SO_2 and its tendency to deposit on indoor surfaces (Yocom, 1982). The most important class of indoor source is the kerosene heater. Kerosene contains small but variable amounts of sulfur which when burned may produce measurable indoor concentrations of SO_2 (Leaderer *et al.*, 1984). An important but rare source of indoor SO_2 is a leaky furnace vent. In a study of indoor–outdoor relationships for SO_2 and other pollutants, Yocom *et al.* (1971) measured SO_2 concentrations above 1 ppm in a home with a leaky vent pipe on the coal-fired heating system.

SO_2 is an irritating gas at high concentrations (above about 1 ppm). The principal concern about SO_2 in the ambient atmosphere (including indoors) is as a respiratory irritant, especially for asthmatics and its ability to oxidize to sulfuric acid aerosol and sulfates. SO_2 in the presence of moisture accelerates the deterioration of materials, notably metals and paper (Yocom *et al.*, 1986). The study of the effects of SO_2 and other pollutants on indoor materials is in its infancy.

4.3.2 Available Measurement Methods for SO_2

The U.S. EPA reference method for measurement of SO_2 in outdoor ambient atmospheres is the pararosaniline method (U.S. Code of Federal Regulations, Part 50, 1989). This is a manual method based on passing sample air through a solution of potassium tetrachloromercurate in a bubbler to form a stable complex which is measured colorimetrically after reaction with pararosaniline. Equivalent methods using continuously recording instruments are recognized by the U.S. EPA. One of these is an automated version of the Reference Method, while others are based on physical principles such as flame photometric detection, pulsed ultraviolet fluorescence and second derivative spectroscopy. At present, the two physical principles represent the instruments in common use for outdoor monitoring. Flame photometric detection relies on sulfur-specific emissions from a hydrogen-rich air flame. Since the method does not distinguish between various forms of sulfur, filters are used to remove such interferences as H_2S. Because of the generation of exhaust gases containing water vapor, use of such an instrument indoors may require venting. The method based on pulsed fluorescence measures the intensity of the ultraviolet fluorescence of SO_2 excited by high intensity light.

As seen from inspection of Figure 4.1, the U.S. EPA is not pursuing the development of SO_2 total exposure monitors presumably because much progress has been made since the early 1970s in reducing outdoor SO_2 emissions and because SO_2 is not considered an important indoor pollutant. Nevertheless, as reported by Nagda *et al*. (1987), there are several commercially available portable instruments suitable for indoor air quality monitoring based on pararosaniline chemistry and electrochemistry.

4.3.3 Indoor Concentrations of SO_2

The results of the limited number of studies in which indoor concentrations of SO_2 were measured are summarized in Appendix C. Summary descriptions of these studies are presented by the U.S. EPA (1988a). Most of these studies were carried out in the early days of indoor air quality investigations when the emphasis was on penetration of outdoor pollutants indoors rather than on the effect of pollutants generated indoors on indoor air quality. SO_2 is considered primarily as a pollutant of outdoor generation. The principal findings from these studies are summarized as follows:

- Indoor concentrations of SO_2 were considerably lower than those outdoors. The decay of SO_2 as a result of reaction or deposition on indoor sources and the lack of indoor sources are believed to account for this phenomenon. Indoor decay of SO_2 is affected by the type of interior surface.
- In a few instances, poorly ventilated heating systems burning sulfur containing fuels produced indoor concentrations of SO_2 that exceeded outdoor concentrations.
- Indoor SO_2 studies have been carried out using a variety of SO_2 monitoring systems. Some of the early studies used methods that had serious interferences (e.g., conductiometry); thus, the data from these studies are suspect, but are of historical interest.
- In developed countries where outdoor SO_2 concentrations have been reduced in recent years, indoor SO_2 exposures are believed to be of little importance with regard to possible health effects. However, in certain underdeveloped countries where sulfur-containing fuels such as coal are used in unvented cooking systems, indoor exposure to SO_2 and a variety of other combustion pollutants is a serious health hazard.

4.4 Ozone (O_3)

4.4.1 Sources and Characteristics of O_3

Ozone (O_3) is an air pollutant normally associated with outdoor photochemical smog. Under these conditions O_3 is generated in the lower atmosphere through a complex series of reactions between nitrogen oxides and reactive hydrocarbons in the presence of strong sunlight. Thus outdoor O_3 patterns have a strong diurnal pattern. There are few outdoor sources that generate O_3 directly, but there are a number of indoor sources such as photocopiers, electrostatic air cleaners and to some extent other types of electrical equipment such as motors. O_3 is extremely reactive and one of the most powerful oxidizers in nature. Therefore, without indoor sources, O_3 penetrating from the outdoors decays rapidly. Mueller *et al.* (1973) measured O_3 decay in rooms of several sizes and determined that the first order decay rate is directly related to the surface-to-volume ratio of interior spaces. Humidity increases decay rates with certain types of surfaces, and it was found that the half life in various indoor settings ranged between 5.7 and 11 minutes. Thus, a tightly constructed house will normally provide good protection from exposure to outdoor O_3. Weschler *et al.* (1989) measured indoor and outdoor O_3 concentrations in three office buildings with different air exchange rates. Indoor–outdoor air ratios ranged between 0.2 and 0.8, the highest values occurring in buildings with the highest air exchange rates, approximately 25 ACH.

4.4.2 Available Measurement Methods for O_3

The U.S. EPA Reference Method for O_3 is based on the gas-phase chemi-luminescence from the reaction between ethylene and O_3. Two other methods have been recognized by U.S. EPA as equivalent to the Reference Method. One is based on the chemiluminescence of the reaction between O_3 and rhodamine-B, and the other is based on the ability of O_3 to absorb ultraviolet radiation in which the ultraviolet intensity of two parallel beams are compared, one containing sample air and the other a reference. In the chemiluminescence instruments, small amounts of unreacted gases may be emitted, and those wishing to use such devices indoors, depending on the type of study planned, should determine from the manufacturers if the quantities of these contaminants are sufficient to require outdoor venting of the exhaust. The U.S. EPA considers O_3 to be strictly an outdoor pollutant and is therefore not planning to develop total exposure monitors for O_3. Nagda *et al.* (1987) describe one commercially available portable instrument for O_3 based on the chemiluminescent reaction with ethylene.

4.4.3 Indoor Concentrations of O_3

Like SO_2, relatively few studies of indoor O_3 concentrations have been carried out. The results of the six major studies are summarized in Appendix D, and brief descriptions of the studies are presented by U.S. EPA (1988a). The principal findings from these studies are summarized as follows:

- O_3, a pollutant of outdoor origin and a product of photochemical smog, decays rapidly after penetrating the indoor environment. Indoor concentrations were found to be in the range of 20 to 80% of outdoor concentrations.

- Indoor O_3 concentration patterns tend to follow the typical outdoor diurnal patterns, but peak indoor concentrations lag those measured outdoors.
- There is some indication that mechanical ventilation systems that impart turbulence to indoor air tend to reduce indoor O_3 concentrations.
- Limited data from homes with electrostatic air cleaners indicated that such systems had little or no effect on indoor O_3 concentrations.
- Studies of O_3 in aircraft cabins flying at high altitude showed that in-cabin concentrations are only slightly lower than those outdoors and can reach concentrations that exceed the NAAQS.
- Most of the indoor air quality studies for O_3 used a monitoring device based on chemiluminescence.

4.5 Respirable Particulate Matter (RPM)

4.5.1 Sources and Characteristics of Particulate Matter

While the thrust of this section is on RPM, a brief review of particulate matter of all sizes and its importance will provide useful background.

Airborne particulate matter is ubiquitous and exists in a wide range of particle sizes and chemical characteristics as can be inferred from Table 1.3. Particulate matter is produced by many outdoor and indoor sources. Coarse particles (generally >10 μm in diameter) are produced by mechanical processes such as grinding, abrasion or surface re-entrainment. Particles of this type are most commonly emitted from outdoor sources and consist largely of mineral components (e.g., silicon, calcium, iron, and aluminum) and biological matter (e.g., pollen, insect parts, and vegetation). Such particles do not readily penetrate tightly constructed indoor spaces. Outdoor sources of coarse particles include mineral-based industries, mining operations, fugitive dust from roadways, and agricultural operations. Such large particles are found indoors, but are brought indoors by tracked-in soil and as dust collected on clothing and are air-suspended through various human activities, including floor sweeping.

Inhalable particles (generally less than approximately 10 μm) have both outdoor and indoor sources, and the smallest particles in this category (generally less than approximately 3 μm), termed respirable particulate matter (RPM), are produced principally by vapor condensation and agglomeration of Aitken nuclei (<0.1 μm). Outdoor sources of RPM include combustion processes, metallurgical operations producing metallic fume, and aerosols produced by photochemical smog. The single most important source of indoor RPM is tobacco smoking. The size range of aerosol particles in environmental tobacco smoke (ETS) is generally <1 μm. Other indoor sources include unvented or improperly vented heating systems and cleaning operations such as vacuuming. Indoor RPM consists of complex mixtures of organic compounds (e.g., cigarette and cooking smoke, indoors) and sulfates, nitrates and ammonium salts produced both outdoors and indoors.

The following two examples illustrate some of these factors. Alonza *et al.* (1979) measured indoor–outdoor relationships for a number of elemental components of total filterable particulate matter in various types of interior spaces. They found that indoor–outdoor ratios for calcium and iron were much lower than for zinc, lead, and bromine. Tosteson *et al.* (1982) measured the indoor–outdoor ratios of RPM and its iron, lead and aluminum content and found that they were not far from 1.0 (range: 0.74 to 1.08), indicating that in the size range of RPM each of the metallic components readily penetrated the indoor environment. However, the indoor–outdoor ratio of the RPM was 2.5 indicating significant indoor sources of particles in this size range, hypothesized to be due to ETS.

Until July 1987, outdoor ambient air quality standards for particulate matter in the United States had been based on total suspended particulate matter (TSP). This category of particulate matter is defined by the sampling method, the High Volume Air Sampler (Hi-Vol). This is a large sampler drawing air through an 8 inch × 10 inch (22.9 cm × 25.4 cm) glass fiber filter or other high efficiency filter media at a rate of approximately 1200 LPM, and collects particles over an extremely broad particle size range. Under most conditions, this device has a 50% particle cut size of approximately 30 μm. In other words, half of the particles by weight are above and below this size. It is obvious, therefore, that the results from a Hi-Vol are extremely sensitive to the relative number of large particles present in the air being sampled. The Hi-Vol has been used for many years to provide data for determining compliance with National Ambient Air Quality Standards in the United States (see Table 1.2), and some of the early indoor air quality studies used the Hi-Vol or modifications thereof (Yocom, 1982). The Hi-Vol sampling rate is far too high for use in most indoor sampling locations. Furthermore, the noise generated by such devices may be a problem in indoor sampling applications. Yocom *et al.* (1970) developed and applied an indoor TSP sampler drawing 140 LPM through a 4-inch (10.2 cm) diameter filter located in a "bird house" whose design and capture velocity and particle cut size (30 μm) simulated the Hi-Vol. Results from this system were believed to be comparable to the full-sized Hi-Vol. Other early studies used standard 47 mm diameter open filters (Moschandreas *et al.*, 1978). Because of the sensitivity of these methods to extremely large particles, comparability of data between these early studies cannot be made.

As noted in Table 1.2, the outdoor air quality standard for particulate matter is now based on "inhalable particulate matter" with a 50% cut size of approximately 10 μm (PM$_{10}$). This standard is intended to protect against exposure to particles capable of entering the trachea and lungs and contributing to health effects such as bronchial cancer, bronchitis and emphysema (Miller *et al.*, 1979). Lioy *et al.* (1980) reviewed data from a large number of inhalable particulate matter studies and listed the large number of monitoring devices used in these studies.

There is increasing concern about "respirable particulate matter" (RPM) which is that fraction of ambient particulate matter capable of penetrating through the airways of the lower respiratory tract (tracheobronchial tree) of healthy adults and can deposit in those portions of the lungs (alveoli) not protected by ciliary action (Miller *et al.*, 1979). The 50% cut size for particles in this size category is in the range between approximately 2.5 and 5 μm depending on the lung penetration curve used. The Lioy *et al.* (1980) review includes a listing of a number of devices used for measuring particles in this size range in both outdoor and indoor sampling. Fletcher (1984) summarized data on 19 personal monitoring and portable instruments for monitoring inhalable and respirable particulate matter.

Since little use is now being made of particulate matter samplers for total particulate matter and TSP, the following two sections of this chapter will deal primarily with RPM and PM_{10}.

4.5.2 Available Measurement Methods for RPM

Devices and methods for measuring RPM and other types of particulate matter are legion. Most are applicable in one form or another to indoor sampling and some are feasible personal samplers. One way to distinguish between these devices is to categorize them as samplers or monitors. Samplers collect a sample which must subsequently be analyzed (e.g., weighed) to produce a measure of particulate matter concentrations in the air. Monitors provide a direct reading of particle concentration or a surrogate for the concentration (e.g., light scattering).

Fletcher (1984) described the operating features of 19 commercially available particle samplers and monitors. This review showed that the most prevalent configuration for samplers and some monitors was an inertial pre-separator such as a cyclone or impactor to remove larger particles followed by a filter. The 50% cut size for these devices varied widely. Several devices are based on light scattering or β-attenuation created by the particle deposits on a collecting surface.

Koontz and Nagda (1986) evaluated in the field six different sampling monitoring devices that represented most of the techniques reviewed by Fletcher (1984). The samplers and monitors evaluated are listed in Table 4.1. The following conclusions are drawn from their analysis:

- The National Bureau of Standards sampler shows excellent precision and accuracy but pump-related problems need to be resolved.
- The cyclone has reasonably good precision and accuracy, but its low flow rate requires a longer sampling duration per measurement.
- The MiniRAM^{TM} shows excellent precision but is affected by temperature and tends to overpredict when particulates are mainly in the fine-particle size range.
- The MARPLE Personal Cascade Impactor has shown the most erratic response and seems to overpredict when particles in the coarser size range are encountered.

A widely used personal sampler for RPM is that developed by Harvard University

Table 4.1   Features of particulate monitors/samplers (Koontz and Nagda, 1988)

Monitor/sampler (originator)	Sample collection and/or measurement technique	Nominal 50% particle-size cut point (μm)	Flow rate (liter/min)	Sampling mode
Dichotomous sampler (Sierra-Andersen)	Size-selective inlet followed by virtual impactor with filters for PM_{10}/RPM fractions	10, 2.5	16.7	Stationary
Portable ambient particulate sampler (National Bureau of Standards)	Impactor inlet with series filtration for PM_{10}/RPM fractions	10, 3.5	6.0	Portable
MARPLE personal cascade impactor (Andersen Samplers, Inc.)	Cascade impactor for PM_{10}, RPM and other size fractions	10, 3.5, others*	2.0	Portable/Personal
Cyclone (various originators)	Cyclone followed by filter	3.5	1.7	Portable/Personal
MiniRAM™ particulate monitor[†] (GCA Corporation)[‡]	Optical sensor	None	Passive	Portable/Personal
Handheld aerosol monitor[†] (PPM, Inc.)	Optical sensor	None	Passive	Portable

*   21, 15, 6, 1.6, 0.9, and 0.6 μm.

[†]   Either of these devices can be configured with a cyclone and pump to actively sample the respirable size fraction only.

[‡]   Current name: MIE, Inc.

as part of the Six-Cities Study (Dockery and Spengler, 1977). The sampling train consists of a 10 mm diameter cyclone with a 50% cut size of 3.5 μm for removal of the non-respirable particles followed by a 37 mm Fluoropore filter and a small battery-powered pump operating at 1.7 LPM.

A relatively recently developed monitor for RPM, PM_{10} or TSP permits real-time indoor measurement of any of these forms of particulate matter depending upon the type of sampling head. The device called the TEOM™ collects pre-sized particles on a filter located on the end of a hollow tapered glass sampling tube. As particles accumulate on the filter, the natural frequency of the sampling rod changes. This change is continuously monitored, and the instrument's computer determines real-time particle concentrations. The results of measurements with this device compare favorably with results from manual methods. The U.S. EPA has included this method in a compendium of methods applicable to indoor air sampling (U.S. EPA, 1989). Figure 4.4 is a photograph of this device in its form applicable to indoor air monitoring.

The U.S. EPA (1988a) in reviewing the state-of-the-art in personal and portable samplers and monitors for RPM recommends research to develop new and improved samplers and monitors, for example:

- More sensitive and accurate direct reading RPM monitors based on such phenomena as light scattering and piezoelectric frequency shift.
- A smaller version of the dichotomous sampler to collect both PM_{10} and RPM.
- Development of personal monitors for collection of both PM_{10} and RPM at higher flow rates which will allow shorter sampling periods.

4.5.3 Indoor Concentrations of Respirable Particles

Appendix E presents the summary data on studies which report RPM. Descriptions of these studies are presented by the U.S. EPA (1988a). Because of the extensive research done on a specific source of indoor particles, environmental tobacco smoke, these studies are reported separately in a subsequent section. Several early studies (pre-1977) investigated indoor "smoke," a measure of particle reflectance or light absorbance. Since the relationship between "smoke" and respirable particles is not available, those studies are not reported here. Other early studies using a modified Hi-Vol approach are also not reported as those methods certainly collected particles of larger diameter than the respirable fraction of interest here. Most of the microenvironments reported are residences, and the distinct trend of the results is for indoor concentrations of RPM to be greater than outdoor concentrations.

By far the largest single database for indoor and total exposure to RPM is that developed as part of the Harvard Six Cities Study (Spengler *et al.*, 1981). Other more limited studies by others have dealt primarily with the effect on indoor concentrations of RPM of indoor sources such as smoking and wood stoves. The principal findings from the studies listed in Appendix E are:

- Total exposure to RPM is dominated by indoor concentrations.
- Indoor concentrations of RPM tend to be higher than outdoor concentrations.
- There is some indication that indoor concentration of RPM are higher in the winter than in the summer.
- Smoking and operation of combustion devices indoors increase indoor RPM concentrations.
- Some limited studies in which RPM samples were collected in several rooms of houses showed that there appeared to be good mixing of particles within the houses.

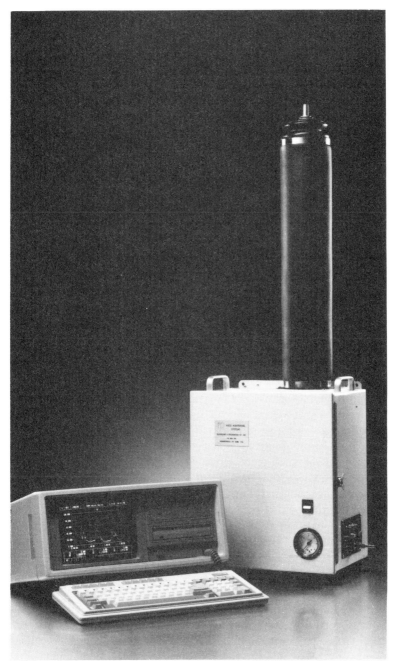

Figure 4.4   The series 1200 TEOM monitor for real-time measurement of indoor particulate matter (U.S. EPA, 1989). (Photograph courtesy of Rupprecht and Patashnick Co., Inc.)

- No clear relationship was found between indoor RPM concentrations and ventilation rates.
- One limited study showed that in homes with smokers the presence and use of an air conditioner tends to increase indoor RPM concentrations.
- Most measurements reported were made with samplers using a cyclone with a cut size of 3.5 μm. Some measurements were made with dichotomous samplers with a 2.5 μm cut size.

4.6 Lead

4.6.1 Sources and Characteristics of Lead (Pb)

Lead-containing particulate matter in the indoor environment has three principal sources:

- Lead-containing particulate matter in the outdoor atmosphere that migrates indoors as part of air infiltration.
- Lead-containing outdoor surface soil that is tracked indoors and becomes air entrained.
- The entrainment of lead-containing particulate matter from the deterioration or mechanical disturbance of indoor lead-base paint.

In the U.S. the outlawing in 1976 of tetraethyl lead as an anti-knock compound in automotive gasoline has gradually but significantly reduced outdoor lead levels and the contribution of outdoor ambient lead concentrations to indoor concentrations. The principal source today of indoor airborne lead is tracked-in soil from areas adjacent to heavily travelled roadways or lead smelters. This effect, of course, applies only to homes located in areas of contaminated soils. Exposure of children to lead in older homes, particularly tenements in larger cities, is a problem related to deteriorating lead-based paint. However, the principal route of entry of the lead is believed to be through ingestion of flaking paint rather that inhalation of lead-containing particulate matter. Lead in drinking water is another potential source of indoor lead exposure, but is outside the scope of this book. Those readers wishing to learn more about lead as an air pollutant are directed to the Criteria Document on lead prepared by the U.S. EPA (1986).

4.6.2 Available Measurement Methods

Since lead in the indoor environment is in particulate form, the methods for its collection are basically the same as those described under respirable particulate matter. The method of collection is generally by filtration or impaction, and the collecting surface must, of course, be compatible with the analytical method for lead determination. The common methods for measuring the lead content of collected particulate matter are atomic absorption, X-ray fluorescence and wet chemical methods.

4.6.3 Indoor Concentrations of Lead

Indoor concentrations of particulate lead have been measured in a significant number of indoor air quality studies. In some studies, the Pb content of total particulate matter was determined, while in other studies only RPM was considered. These studies are summarized in Appendix F, and a more detailed analysis of each study has been presented by the U.S. EPA (1988a). The principal findings from a review of these studies are summarized as follows:

- In the U.S. the conversion to unleaded fuels for motor vehicles fuels has gradually lowered indoor and outdoor ambient concentrations of lead. Only the earliest indoor air quality studies in the early 1970s showed any indoor concentrations exceeding the current NAAQS (1.5 μg/m^3 averaged over 3 months).
- Indoor concentrations of lead are usually less than concentrations outdoors confirming the outdoor source of lead-containing particulate matter. However, indoor–outdoor ratios for lead in which total particulate matter was measured tend to be higher (closer to 1.0) than the indoor–outdoor ratios of total particulate matter and elements associated with crustal materials such as calcium. This difference is believed to be associated with the small size of lead particles and their greater ability to penetrate indoor spaces.
- While indoor exposures to lead in the home tend to be less than those immediately outside the home, studies in which total exposure to lead were measured with personal RPM monitors showed that total exposure is often greater than would occur in the home or outdoors near the home because of exposure received during vehicle travel.
- In areas where outdoor soil is contaminated with lead (e.g., in areas near lead smelters), indoor lead concentrations are strongly influenced by reentrained floor dust.

4.7 Radon and Decay Products

4.7.1 Sources and Characteristics of Radon (Rn)

Radon, as opposed to other indoor pollutants, is not of anthropogenic origin. Its source in the biosphere is primordial radioactive elements, their decay products (daughters or progeny), and cosmic radiation. Radon is a noble gas, occurring as part of the natural uranium decay chain in soil and rock. Outdoors, radon is dispersed by winds and vertical mixing; thus it is prevented from accumulating to significant concentrations. Enclosed microenvironments, particularly residential structures, present exposures to radon. In general, soil gas is the primary source of radon in residences. Other important sources of radon are earth-based building materials and domestic water from wells. Building materials such as alum-shale concrete, used in Sweden, may have radon concentrations two orders of magnitude higher than other materials. Granite, shales, and phosphate-containing soils in the United States also contain elevated levels of radon.

Because sources of indoor radon are varied and are affected by several variables, it is useful to review these items in more depth than is necessary for the other pollutants presented in this chapter. A more detailed discussion of this topic is presented by Nero (1983a) and Nazaroff and Nero (1984 and 1988).

In considering building materials as radon sources, both the radionuclide concentration and the diffusion (emanation) characteristics of material must be determined. Ingersoll (1983) reported that in 100 concrete samples from 10 U.S. cities, radon yields ranged from 0.25 to 1.95 pCi/kg/h; however, he concluded that these materials are not a primary source of radon in typical U.S. houses. Nero (1983a) points out that in larger structures (e.g., apartment buildings) the building materials may contribute a greater share of the source strength. However, Nazaroff and Nero (1984) note that only in the absence of other sources of radon do normal building materials become important as a radon source in residential structures. Materials known to be significant contributors are alum-shale concrete and phosphogypsum.

Well water is another source of indoor radon. Concentrations are regionally dependent, related to local geology. Hess *et al.* (1982) determined the component of indoor radon associated with water usage in 100 houses in Maine. Indoor air concentrations ranged from 0.05 to 135 pCi/liter. Regression analyses of these data show that 10000 pCi/liter of radon in water contributes 1.07 pCi/liter of radon to the air. According to Sachs *et al.* (1982), water has the greatest variability of all three radon sources. "Public" or "city" water taken from surface supplies or stored in reservoirs has low radon levels. Bruno (1983) and Kothari (1984) have found that radon in well water occurs in only a few regions of the United States.

It is generally accepted that soil is the predominant source of radon indoors (Bruno, 1983; Hildingson, 1982; Nero, 1985). Nazaroff and Nero (1984) state that soil is the dominant source because homes without radon-containing building materials still contain elevated levels of radon, and flow-inducing mechanisms, i.e., infiltration, can enhance radon transport from the ground into the structure. Thus, infiltration, through small pressure differences between the lower part of the house interior and the outdoors, appears to be a driving force promoting the flow of radon out of the soil and into the house.

As soil is usually the dominant source of radon, it follows that certain characteristics of homes can enhance this effect. According to Nazaroff and Nero (1984), building substructures can be ranked from greatest enhancement of soil emissions to least enhancement; from full basements to crawl space to slab-on-grade.

Geological conditions have not been shown to be useful *a priori* predictors of radon levels in homes since the major parent element, uranium, is not distributed in any simple way with geological conditions. However, in a given area, once homes with high radon levels are known, mapping of the local geology can help identify other potentially high radon homes. Thus, geographic location is a strong determinant of indoor radon concentration, but its predictive value is limited. To predict indoor concentrations such variables as housing substructure, ventilation patterns, and soil permeability must be known. Sextro *et al*. (1987) have shown that soil radium is not the only predictor of indoor radon. They report that elevated indoor concentrations of radon occur if the soil is highly permeable and the house extracts a large fraction of its makeup air via the basement.

4.7.2 Mechanism of Toxicity

The lung is considered the primary target organ for radon toxicity because radon decay products attach to inhalable particles which become deposited in the lung, especially in the tracheo-bronchial tree (Nero, 1983a). Natural sources of radiation can also affect other tissues of the body. The short-lived decay products of radon, i.e., polonium, lead, and bismuth, are chemically active and are collected in the lungs. The most significant dose arises from α-decay of polonium isotopes (Nero, 1983b). These isotopes have short half-lives (less than 30 minutes), and once they are collected in the lung, they decay to lead-210 and irradiate the surrounding tissue before the body's lung-clearance mechanisms remove them. This causes large radiation doses in the small volume of the bronchial epithelium (Martell, 1984). The decay sequence of radon-222 to lead-206 is shown in Figure 4.5. For future reference, note that ^{218}Po, ^{214}Pb, and ^{214}Bi are also referred to as RaA, RaB, and RaC, respectively.

The radon progeny are produced as individual atoms which rapidly form clusters with water, oxygen, or other gases as a result of the electric charge they acquire upon decay. Some of these clusters may subsequently become attached to aerosol particles. Because the respiratory tract deposition sites are dependent upon the characteristics of the aerosol particle (size, mass, aerodynamic shape), there are differences between the dose imparted to the body by the clusters and the particles, and to some extent, between the particles of different sizes and shapes. While the relative fractions of the attached and unattached radon daughters are important from a health-effects standpoint, they are rarely reported. Attached progeny tend to deposit where they are carried by the particles to which they are attached (James, 1987). Because of their efficient deposition in the lung and their higher dose value per unit of potential alpha energy than progeny in the unattached state, the toxicity of the attached fraction is significant. The unattached fraction increases with decreasing aerosol concentration and with reduced residence time (i.e., increased pulmonary ventilation rate). The attached/unattached ratio can vary appreciably between the much dustier mines from which the epidemiological data have been obtained and residential environments. James (1987) indicates that dose per unit exposure in homes is marginally higher than in mines.

Concentrations of radon progeny in the air are usually expressed in picocuries per liter (pCi/liter) or analogous units (that is, activity per unit volume). Another way to

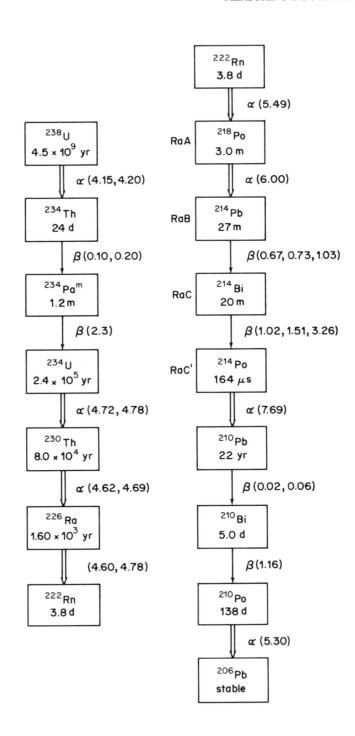

express radon concentration is as becquerel per cubic meter (Bq/m^3). A becquerel is a unit of activity; 1 disintegration per second equals 1 becquerel. For conversion, 1 pCi/liter equals 37 Bq/m^3. However, this quantity does not distinguish between the decay products releasing the activity, and thus does not give a true estimate of the total alpha energy potentially released as the progeny further decay to lead-210. Since alpha energy is associated with lung cancer, exposure to radon progeny is usually described in terms of working levels (WL), where 1 WL = 1.3 $\times$ 10^5 MeV/l, which corresponds to air containing approximately 100 pCi/liter of each of the short-lived daughters (RaA, RaB, and RaC). Exposures are also reported in working level months (WLM), where 1 WLM = 1 WL $\times$ 173 hours.

Although the progeny are usually measured as a single unit (total activity), one should know the relative amounts of each to compute the exposure, in WL. The actual formula is:

$$WL = 0.00103 \ C_{RaA} + 0.00507 \ C_{RaB} + 0.00373 \ C_{RaC}$$
(Radon progeny concentration in units of pCi/liter)

Since the progeny are in equilibrium with each other only in a totally closed system without loss through deposition, one must assume a reasonable "equilibrium factor" (F), to describe the relative ratios of the progeny. This factor is the ratio of the actual progeny energy concentration to the energy concentration one would expect if each progeny had the same activity level as the radon present. Nero (1985) reports residential equilibrium factors ranging from 0.3 to 0.7. According to James (1987), there is an inverse relationship between equilibrium factor and unattached fraction. Thus, as the unattached fraction decreases, the equilibrium factor approaches 1.0. Based on the inhalation dose model developed by James (1987), lung dose is proportional to the radon gas concentration over a wide range of conditions. Therefore, monitoring data simply for radon and not for unattached progeny can be used to estimate dose with reasonable certainty.

The equilibrium factor, F, is affected by the removal processes occurring in a home. Radon progeny can be removed through ventilation, plateout, or a mechanical filtration. Elevated ventilation rates will result in a lower value of F. Similarly, increasing air movement in a room will serve to improve the plateout rate and thus produce a lower equilibrium factor. The use of air cleaning devices (electrostatic precipitators or filters) can also provide a removal path, thus lowering the value of F.

Given $F = 0.5$, the exposure rate for a total progeny activity concentration of 1 pCi/liter is:

$$\frac{1}{2} \times \frac{pCi}{liter} \times \frac{1 \ WL}{100 \ pCi/liter} \times \frac{1 \ WLM}{(1 \ WL)(173 \ h)} \times \frac{8760 \ h}{year} = 0.25 \ \frac{WLM}{year}$$

Figure 4.5 Radon-222 decay sequence showing half-lives and type of decay. (Reproduced from Nero (1983b) with permission of the Health Physics Society)

The actual exposures can be adjusted for different values of F, if more knowledge of the factor is known. However, this "conversion" formula, 1 pCi/liter = 0.25 WLM/yr, provides a good approximation (assuming, of course, that the individual in question spends the bulk of the time indoors).

To convert directly from activity concentration to WL, one uses only the first three terms of the expression above:

$$1 \text{ pCi/liter} = 0.005 \text{ WL}, \quad \text{where } F = 0.5.$$

4.7.3 Available Measurement Methods for Radon and Radon Progeny

The general techniques for measuring radon and radon progeny are summarized below. Radon and the progeny exist in a relatively constant equilibrium concentration in residential settings (George, 1986; Nero, 1985). Radon progeny are the most difficult to monitor; thus direct measurement of radon is generally preferred for survey applications. Further discussion of the methods is presented by George (1986). A publication by the U.S. Environmental Protection Agency (U.S. EPA, 1986) provides a detailed discussion of several monitoring techniques along with information for field use and quality control procedures. The U.S. EPA (1986) reference is important as it provides standardized protocols for each method with the objective of generating reproducible measurements with limited variability. The methods which U.S. EPA (1986) has evaluated and provided standardized procedures for radon are: continuous radon monitors (using scintillation cells), charcoal canisters, alpha-track detectors, grab sampling; and for radon progeny: continuous working level monitors, radon progeny integrating sampling units (RPISU), and grab radon decay products (using filter collection). Another resource for two radon methods, filter paper collection and alpha activity measurement and the charcoal canister, can be found in Lodge (1988).

The methods summarized below are categorized according to species measured: radon and radon progeny. Within each of these two groups, methods vary according to grab, continuous, and integrated techniques. All vary in cost with the integrated radon techniques being the least expensive and the continuous radon progeny techniques being among the most expensive. Tables 4.2 and 4.3, adapted from George (1986), provide a concise summary of the various methods for measuring radon and radon progeny, respectively. The last column indicates if EPA has developed measurement protocols for that method. In general, grab sampling techniques are used for screening purposes or diagnostic studies to identify routes of entry for radon soil gas. Continuous sampling is required for in-depth studies where real-time fluctuations in concentration are of interest. Integrated techniques provide information on average radon concentrations over several days to months. Because of the long sampling time, these latter techniques have fairly low detection limits.

Table 4.2 Instruments and methods for measuring radon in air[*]

Instrument and method	Sampling application	Principle of operation	Sensitivity	EPA measurement protocol
Scintillation cell	Grab	Scintillation alpha counting.	<0.01–1.0 pCi/liter	Yes
Ionization chamber	Grab (laboratory only)	Sample transferred into ion chamber. Pulse of current counting.	<0.005 pCi/liter	No
Active continuous scintillation	Continuous	Flow through scintillation cell alpha counting.	<0.1–1.0 pCi/liter	Yes
Passive diffusion electrostatic monitor	Continuous	Radon diffusion into sensitive volume. ^{218}Po collected on scintillation detector electrostatically.	0.5 pCi/liter for 10-minute counting intervals	No
Passive diffusion radon only monitor	Continuous	Radon diffusion into sensitive volume. Radon progeny removed. Alpha particles from radon decay only with alpha scintillation counter.	0.1 pCi/liter for 60 minute counting intervals	No
Passive track etch monitor	Integrating	Alpha sensitive film registers tracks when etched in NaOH.	0.2, 0.4 pCi/liter - month depending on size	Yes
Passive activated carbon monitor	Integrating	Radon adsorption on activated carbon. Gamma counting with gamma analyzer for ^{214}Pb and ^{214}Bi.	0.2 pCi/liter for 100-hour exposure	Yes
Passive electrostatic-thermoluminescence monitor	Integrating	Radon diffusion into sensitive volume. ^{218}Po collects on thermoluminescence detector electrostatically.	0.03–0.3 pCi/liter (depending on size of diffusion volume) for 170-hour exposure	No

[*] *Source:* A. George, "Instruments and Methods for Measuring Indoor Radon and Radon Progeny Concentrations," in *Indoor Radon, Proceedings of an APCA International Specialty Conference, Philadelphia, PA,* 1986. See original citation for references. Reproduced with permission of the Air and Waste Management Association.

Table 4.3 Instruments and methods for measuring radon progeny in air*

Instrument and method	Sampling application	Principle of operation	Sensitivity	EPA measurement protocol
Kusnetz-Rolle	Grab sampling for WL	Collect sample on filter for 5–10 min. with alpha counting.	0.005 WL	Yes
Tsivolglou and modifications[†]	Grab sampling for individual radon progeny and WL	Collect sample on filter for 5–10 min. with alpha counting.	0.1 pCi/liter each of ^{218}Po, ^{214}Pb, ^{214}Bi and 0.0005 WL	Yes
Tsivolglou and modifications	Continuous - instant radon progeny and WL monitoring	Collect sample on filter for 2–3 min. with alpha and beta counting.	0.1–1.0 pCi/liter 0.001–0.01 WL depending on flow rate	Yes
Radon progeny integrating sampling unit (RPISU)	Integrating	Collect sample on filter for 1–2 weeks Detect with thermoluminescence material (dyprosium activated CaF$_2$).	0.0001 WL	Yes
Thermoluminescence modified WL monitor	Integrating	Collect sample on filter for 1–2 weeks. Detect with thermoluminescence material (LiF).	0.0005 WL	No
Surface barrier WL monitor	Integrating	Collect sample on filter continuously. Detect alpha radioactivity with silicon surface barrier detector.	0.00005–0.005 WL depending on flow rate rate	No
Radon/Thoron WL monitor	Integrating	Collect sample on filter continuously. Detect radon and thoron daughter radioactivity with alpha-sensitive film.	0.001 WL in 240 h	No

* Source: A. George, "Instruments and Methods for Measuring Indoor Radon and Radon Progeny Concentrations," in Indoor Radon, Proceedings of an APCA International Specialty Conference, Philadelphia, PA, 1986. See original citation for references. Reproduced with permission of the Air and Waste Management Association.
† The method reported by Lodge (1988) is a variation on this technique.

Radon.   There are two techniques for continuously monitoring radon: the continuous scintillation cell and the diffusion electrostatic monitor. The scintillation cell instrument filters the air and draws it into the scintillation cell. Scintillation cells are containers of variable size, ranging from 0.09 to 2 liters. They have a transparent bottom, coated with silver-activated zinc sulfide phosphor. The cell is attached to a photomultiplier tube. Air is drawn into the cell (either via a pump or by evacuation), and as the radon decays it emits alpha particles, which when they disintegrate create light pulses (scintillations) from the interaction with the coated surface. The number of pulses is proportional to the radon concentration. According to George (1986) scintillation cells will last for several years, if properly maintained. The collection devices, as long as they are impervious to radon, can range from metal containers to Tedlar^{TM} bags. Samples are transferred to scintillation cells and counted. Counting is frequently done on a laboratory unit. The diffusion electrostatic monitor is a passive device with a scintillation detection system. Air enters the sensing cell by molecular diffusion. These instruments range in sensitivity from 0.01 pCi/liter to 1 pCi/liter.

The most commonly employed methods for monitoring radon are integrated, passive techniques. The two types are charcoal canisters and alpha-track detectors. The charcoal technique uses canisters which contain activated charcoal. Radon passively diffuses into the canister. Radon concentrations are determined by counting the gamma particle emissions emitted during the decay series. This is done on a sodium iodide detector with a multichannel analyzer. The canisters can be exposed for periods of 4 to 7 days with analytical sensitivities of 0.5 pCi/liter. The alpha-track detector contains a small piece of sensitive plastic film in a filter covered container. The filter prevents particles and radon progeny from depositing on the film. The alpha particles from the radon which diffused across the filter leave traces on the film. The film is subsequently developed and the tracks (or traces) counted. Exposure times are generally on the order of 1 to 2 months. Sensitivity varies according to the number of tracks counted, but can range from 0.4 pCi/liter to greater than 1 pCi/liter.

The grab sampling technique for radon involves collection of air in a flask coated with zinc sulfide phosphor. Alpha particles create scintillations which are counted with a detector and photomultiplier tube. Sensitivities depend on the air volume sampled, with a lower bound of approximately 0.1 pCi/liter.

Radon progeny.   Radon progeny monitors were originally developed to monitor working levels (WL) in areas known to be contaminated with uranium tailings (George, 1986); thus they are generally active monitors. The continuous techniques use filtration with subsequent counting of alpha particles. One method, the continuous working level monitor, draws air through a filter cartridge at a low flow rate, and the alpha particles are counted as they decay. A sampling period of 24 hours is recommended by the U.S. EPA (1986). The other method, Radon Progeny Integrating Sampling Unit (RPISU), draws air continuously through a detector assembly. Radon progeny are collected on a membrane filter which is located on a thermoluminescence detector. The sampling time is 3 to 7 days, and sensitivities of approximately 0.0001 WL have been reported by George (1986).

Grab sampling for radon progeny usually consists of collecting the decay products from a known volume of air on a filter and immediately counting the activity on the filter following collection. Although there are several methods available, U.S. EPA (1986) presents two as examples. The Kusnetz procedure provides information on working levels but not on individual progeny concentrations. Sampling times are on the order of 5 minutes and filters are counted for total alpha activity. The Tsivoglou procedure, as modified by Thomas (U.S. EPA, 1986), follows the same basic principle; however, by counting the same filter at different intervals concentrations of the three progeny are determined. The sensitivity of these techniques is approximately 0.0005 WL.

4.7.4 Indoor Concentrations of Radon

Radon studies in the literature are far too numerous to comment on individually. Therefore this discussion presents results of some of the larger studies which have been conducted as well as highlights from extensive review articles. Appendix G presents a table developed from a review article by Nero *et al.* (1986) which represents the most complete compilation yet of U.S. radon studies. Many of the studies included in Appendix G are summarized below. In reviewing the literature on indoor radon concentrations, one should be aware of two aspects of the study: its purpose and the manner in which sampling locations (i.e., homes) were selected. Frequently, homes are not randomly selected, as a particular phenomenon is of interest to the researcher. Such data should not be used to extrapolate to a large distribution of homes.

The U.S. EPA has established a guideline concentration of 4 pCi/liter, above which remedial action is recommended. EPA also provides guidance on the selection of remediation measures. The most current information may be obtained from the Office of Radiation Programs, Washington, DC.

Harley (1981) summarized the indoor measurements of radon-222 concentrations in different countries and in different cities. Much of the data, however, were collected at locations where elevated levels were expected and cannot be readily applied to other, or entire, regions.

Abu-Jarad and Fremlin (1982), studying two high-rise buildings in Birmingham, England, found the highest radon levels in basements, followed by those of the first floor. However, there appeared to be little difference between levels measured on floors 2–17, an observation attributed to uniform mixing in the ventilation system.

Hess *et al.* (1982) measured radon at 5 locations (4 indoors, 1 outdoors) in 100 homes in Maine between October, 1980 and May, 1981. Radon concentrations in water were determined for each home. The component of indoor air radon associated with the water source was found to be 0.8 pCi/liter per nCi/liter in water at 1.0 ACH. Airborne radon concentrations on the second floor were lower than those on the first floor by approximately a factor of two.

Moschandreas and Rector (1982) reported results from radon grab samples in the living rooms and basements of 50–60 residences in the Washington, DC area. All homes were closed up with heating, ventilation, and air conditioning systems turned off for 8 hours prior to sample collection. The houses were grouped geographically, in reference to their proximity to an experimental residence, as neighborhood, town, and rural. In all areas, the basements had consistently higher radon concentrations than the main floor areas. The authors found no systematic relationship between radon levels and infiltration, so the differences in concentration appear to result from the variability in the source strengths.

Prichard *et al.* (1982) reported radon concentrations determined at several locations in each of approximately 200 dwellings (100 in the Houston, Texas area and 100 in central Maine) over 3–5 months averaging periods during the cooling (in Texas) and heating (Maine) seasons. The distribution of concentrations was log normal; the geometric mean in Maine homes was 1.5 pCi/liter and in Houston homes was 0.47 pCi/liter. The higher radon levels in Maine appear to be related to both the water and soil sources of radon. Differences in construction methods (e.g., more slab foundations in Texas than in Maine) also contributed to the higher Maine levels.

Indoor radon concentrations were monitored as part of a study of air exchange rates and indoor air quality conducted in Rochester, NY (Offermann *et al.*, 1982). The sample of 58 occupied homes included mostly low-infiltration housing. For a 1-week sampling period, radon concentrations ranged from less than detectable to 2.2 pCi/liter and were not correlated with air exchange rates. Based on these and HCHO and NO_2 measurements the authors concluded that when contaminant source strengths are low, acceptable indoor air quality can be maintained, even in energy efficient homes.

Gesell (1983) reviewed studies on "background" indoor radon concentrations conducted in areas without high soil source emissions. In the United States, geometric mean concentrations ranged from 0.77 to 1.7 pCi/liter.

Nero *et al.* (1986) compiled from numerous primary publications a survey of radon concentrations in 1700 single family homes in the U.S. (see Appendix G). In their analysis the authors assert that the data are log normally distributed and that the geometric mean is the best representation of central tendency. The authors conclude that the distribution of radon concentrations is determined by variations in source strength and that ventilation has less influence than the source strength. Accounting for differences in sampling technique and season, the annual average radon-222 concentration is 1.5±0.2 pCi/liter and the geometric mean is 0.9±0.1 pCi/liter. The authors estimate approximately 1–3% of homes in the U.S. exceed 8 pCi/liter.

4.8 Formaldehyde

4.8.1 Sources and Characteristics of Formaldehyde (HCHO)

There are isolated industrial sources of formaldehyde, but its principal outdoor source in urban areas is photochemical smog. Indoors, formaldehyde, a common component in synthetic resins, is emitted from a variety of sources, including particle-board, glues and resins in furniture, carpets and paneling, urea-formaldehyde foam insulation (UFFI), various treated fabrics (e.g., "permanent press" material), and environmental tobacco smoke. Mobile homes represent a distinct type of microenvironment associated with levels of HCHO that frequently are much higher than for other residential settings. Mobile homes are usually constructed with synthetic materials, which often contain urea-formaldehyde resins.

Indoor measurements of HCHO exceed outdoor levels in virtually all studies, with mobile homes having the highest concentrations. In general, UFFI homes have higher concentrations than non-UFFI homes. Elevated levels in non-UFFI residences are often attributable to HCHO offgassing from the structure's building materials, especially particle-board. For these sources of HCHO, an inverse relationship frequently exists between age of source (or home) and indoor HCHO levels. For example, in conventional homes more than five years old, mean concentrations are often below 0.05 ppm. In other types of homes (mobile, UFFI, and new) mean levels can range from 0.1 to 0.4 ppm. Concentrations in new office buildings may also be in this range.

Formaldehyde emissions, and thus indoor concentrations, show significant temporal fluctuations. This variability has been investigated by Meyer (1983) and Wanner and Kuhn (1984) who found that emissions are dependent not only on age but also on temperature and on relative humidity. The Consensus Workshop (1984) noted studies where fluctuations inside houses had diurnal variations up to twofold and seasonal variations up to tenfold, with the highest concentrations associated with higher temperature of the source material.

4.8.2 Available Measurement Methods

There are three commonly employed measurement techniques for HCHO: impinger sample collection with colorimetric analysis, automated wet chemistry, and passive dosimeters. The impinger methods may use two different trapping solutions: distilled water or 1% sodium bisulfite. Both solutions should be cooled to approximately 5 °C for sample collection and storage. The analytical methods are different for each collection media. The 1% sodium bisulfite is analyzed using chromotropic acid according to a modification of the P & CAM 125 (NIOSH, 1977) method; the distilled water solution is analyzed with a pararosaniline procedure (Miksch et al., 1981). Eckmann et al. (1982) includes a description of a specially designed refrigerated unit for housing the sampling trains developed by Lawrence Berkeley Laboratory. Impingers using distilled water should always be cooled, as the HCHO will decompose at room temperatures. Both impinger methods utilize two impingers in series to increase collection efficiency. In the laboratory, the impingers may be analyzed separately or combined. Eckmann et al. (1982) conducted an experiment which compared these methods and found that SO_2 and cigarette smoke could be interferences; however, under laboratory conditions, the methods were equivalent.

An automated wet chemical device, manufactured by CEA Instruments, Inc., Emerson, New Jersey, uses sodium tetrachloromercurate with sodium sulfate as the trapping solution, and pararosaniline is added to produce a colorimetric reaction. It has a lower detectable limit of 0.002 ppm and an adjustable range up to 10 ppm. Nagda et al. (1987) provide a thorough listing of the specifications of this instrument.

Kelly et al. (1989) evaluated two prototype HCHO monitors designed to measure concentrations in the ppb range and below. One of the devices is based on gas phase fluorescence of HCHO and is a modification of a commercially available monitor designed for measuring SO_2. The limit of detection for this method is <100 ppb and should be adequate to detect HCHO concentrations normally found in "problem" indoor environments. The second prototype evaluated is based on an improvement of a design based on selectively wet scrubbing HCHO from the air stream followed by subsequent analytical procedures resulting in quantification of HCHO by fluorescence. The detection limit of this device is 0.2 ppb.

There are three commercially available passive dosimeters for HCHO. The principle of operation is molecular diffusion and sorption. Each manufacturer (Air Quality Research, Berkeley, California; Du Pont, Kennett Square, Pennsylvania; and 3M, St. Paul, Minnesota) employs slightly different analytical methods for the sorption media, but all rely on spectrophotometric quantification of HCHO. These badges are used for integrated samples of up to 1 week of exposure. Shorter sampling intervals are possible; however, the trade off is less sensitivity. With 1 week exposures, the lower detectable limits are 0.010 ppm for both the Air Quality Research and Du Pont badges, and 0.005 for the 3M badge. Further specifications for each of these badges is provided by Nagda et al. (1987).

ASTM Committee D22.05 has developed a passive sampler standard method for formaldehyde (ASTM, 1989). The collection media is an aqueous solution of 0.05% 3-methyl- 2-benzothiazolinone hydrazone hydrochloride (MBTH) in a glass vial (70 mm × 20 mm). The recommended sampling time is 15 minutes to 8 hours. It is sensitive to concentrations ranging from 10 $\mu g/m^3$ (8 ppb) to 10 mg/m^3 (8 ppm). The trapping solution is reacted with a sulfamic acid solution, and the resulting blue cationic dye is measured colorimetrically.

Levin et al. (1989) have reported the use of a passive sampler consisting of a 37 mm filter impregnated with 2,4-dinitrophenylhydrazine and phosphoric acid. It has a reported range of 6–200 ppb; an 8-hour sample has a detection limit of 3 ppb. HCHO is extracted with acetonitrile and analyzed by HPLC with an ultraviolet detector.

In addition to the passive badges, there are two active sampling techniques which use solid sorbents coated with 2,4-dinitrophenylhydrazine. This technique can be used as a personal monitor or as a fixed sampler. Both methods utilize high-performance liquid chromatography for analysis, thus laboratory costs are high. The two methods are reported by Lipari and Swain (1985) and Tejada (1986). Both have sensitivities in the range of 2 ppb-hours. According to the U.S. EPA (1988a) both techniques had blank problems; however, these problems appear solvable in the near future. Because these techniques are relatively recent, as of this writing they have not had widespread application in the indoor air quality field.

4.8.3 Indoor Concentrations of HCHO

There is a large database on indoor concentrations of HCHO. Appendix H summarizes the results of some of the major studies. Since mobile homes are a distinct type of microenvironment and tend to have higher concentrations of indoor HCHO than other types of housing, results from mobile home studies are grouped separately. In spite of the large database, most studies tend to be directed at specific problems such as mobile homes or homes with UFFI and therefore do not provide the basis for an accurate assessment of total population exposure to indoor HCHO. A review of each of the studies summarized in Appendix H has been presented by the U.S. EPA (1988a). The principal findings from this review are summarized as follows:

- Indoor concentrations of HCHO vary widely from below the detection limit to a high of several ppm depending on the type and age of structure, indoor sources, sampling method and its sensitivity, ventilation rate, and sampling duration.
- Concentrations of HCHO in mobile homes, with their extensive use of panelling containing urea-formaldehyde resins and adhesives, are higher than concentrations found in other types of housing.
- In homes (including mobile homes) where HCHO measurements were carried out over time, concentrations decreased with time.
- Smoking and use of indoor combustion appliances appear to have little influence on indoor HCHO concentrations.
- Although one would expect indoor HCHO concentrations to increase with indoor temperature, currently available studies indicate that over the range of normal indoor temperature fluctuations, there is little, if any, influence of indoor temperatures on indoor HCHO concentrations.
- The few studies in mechanically ventilated buildings indicate that HCHO concentrations in such settings are quite low.

- As indicated in Appendix H most of the studies use the impinger sampling method and absorption by chromotropic acid solution. Several studies used passive samplers and one study reported using colorimetric detector tubes.

4.9 Volatile Organic Compounds (VOCs)

4.9.1 Sources and Characteristics

Volatile organic compounds (VOCs) represent a large class of compounds with numerous sources both indoors and outdoors. Exposure to VOCs occurs in many microenvironments associated with human activity. Indoor sources include household products, building materials, drinking water, gas stations, dry cleaners, automobile exhaust and occupant activities such as hobbies or tobacco smoking. Sources of VOCs in the outdoor ambient air include motor vehicle exhaust and various industrial processes such as petrochemical manufacturing and product storage. As yet, there are no U.S. federal standards for any type of VOC in the non-occupational indoor environment.

A distinct class of pollutants as compared to those discussed previously, VOCs are a heterogeneous group of compounds (one study reported 45 different compounds or isomers), with unique sources in different microenvironments. There is no standard method for collecting and analyzing samples and reporting results. VOCs can be measured on site with portable gas chromatographs producing "real-time" data, but this method has limited application when complex mixtures are present in low concentration. The more common method involves three basic steps: adsorption of the vapor phase organics onto a solid sorbent (e.g., Tenax GC^{TM} or activated charcoal), desorption of the organics (either thermally or with a solvent), and analyses of the desorbed compounds (gas chromatography with mass spectroscopy, electron capture, or flame ionization detectors). The specific VOC reported in any research is a function of all these analytical variables. In addition, trapping efficiencies and breakthrough volumes vary among the solid sorbents. Therefore, it is difficult to confirm the accuracy of results since information is sparse on method comparisons.

A further problem which makes it difficult to compare results across studies is the manner in which results are reported. VOC concentrations are highly variable in time and space, often ranging over one or more orders of magnitude. Measures of central tendency (e.g., mean, median, geometric mean) chosen to represent these data are not consistent across studies, and comparing ranges of concentrations is cumbersome and often of questionable value. Other confounding variables in these studies include climate and age of source materials. Although the effect of the latter has been reasonably documented, seasonal variation in VOCs has not been adequately reported.

4.9.2 Measurement Methods

Volatile organic compounds generally are classified as those compounds with saturation vapor pressures at 25 °C greater than 10^{-1} mmHg (U.S. EPA, 1988b). Semi-volatile organic compounds are those which are too volatile to be collected quantitatively by filtration air sampling but not volatile enough for complete thermal desorption form solid sorbents. Generally, these compounds have saturation vapor pressures at 25 °C between 10^{-1} and 10^{-7} mmHg.

Instrumentation and methods for monitoring VOCs must meet several criteria. These include high sensitivity (in the ppb–ppt range), easy portability in the field, and ability to provide accurate and reproducible results. Although there are a variety of methods which meet these criteria, the ones which have found greatest application in the indoor air field are those that involve sample collection in the field with subsequent analysis in the laboratory. Alternative approaches include portable gas chromatographs (GC) with various types of detectors, including the mass spectrophotometer (MS). These sophisticated instruments are expensive to operate and require highly skilled field personnel. Discussion of these laboratory analytical techniques is considered outside the scope of this book.

The following discussion focuses on two different collection techniques: solid sorbents and treated stainless steel canisters. Two different types of solid sorbents are commonly used: polymeric resin or activated carbon. Table 4.4 summarizes the different collection methods, analytes, and analytical parameters. There are three basic steps involved in the use of solid sorbents: adsorption of the vapor phase compounds, desorption of the sorbed material, and analysis of the desorbed material. The stainless steel canisters only require two steps: collection of air in the vessel and direct analysis of the sample.

Table 4.4   Summary of volatile organic compound methods

Collection method	Class of analytes	Desorption	Analytical technique
Polymeric resin (e.g., Tenax GC™)	Volatile non-polar organics having boiling points of 80–200 °C	Thermal	GC*/MS
Activated carbon	Polar and non-polar organics having boiling points of 0–300 °C	Solvent	GC/FID, ECD, NPD
Treated stainless steel canister	Volatile, non-polar organics including aromatics and chlorinated hydrocarbons	None	GC/MS GC/FIC, ECD, PID

GC, Gas chromatography; MS, Mass spectroscopy; FID, Flame ionization detector; ECD, Electron capture detector; PID, Photo ionization detector; NPD, Nitrogen-phosphorus detector.
* With capillary column.

Active and passive sampling methods exist for two of the solid sorbents: Tenax GC^{TM} and activated carbon. In the following presentation, there is a discussion of an issue common to all solid sorbents: breakthrough volume. Next, all the active sampling methods are presented and then the passive techniques.

Active Sampling Techniques

All of the solid sorbent techniques are limited by breakthrough volume. Functionally, this is the saturation capacity of the sorbent allowing elution to occur during sampling. It is defined as the volume of air sampled at which more than 50% of a compound entering the sampling cartridge is stripped off and is lost in the exit stream. Breakthrough volumes are chemical and sorbent specific, and usually are determined experimentally. Once the critical chemical is identified, sample volumes and size of sorbent trap can be modified to provide optimum sensitivity. Most sampling cartridges contain a front and a backup section of sorbent. The two sections are separated by a plug which can be either glass wool or urethane foam. By analyzing these sections separately, it can be determined analytically if breakthrough occurred. Specific types of solid sorbents are highlighted below. As this area of indoor air monitoring is evolving rapidly, the researcher should consult the most current literature on advances in solid sorbent media as well as for information on breakthrough volumes.

Polymeric resin.   The common organic polymer absorbents are Tenax GC^{TM}, XAD, and Ambersorb^{TM}. Tenax GC^{TM} (poly(2,6-diphenyl phenylene oxide)) has found the widest usage, being the method used by the U.S. EPA (Wallace, 1987) for the Total Exposure Assessment Methodology (TEAM) study. It is also used in EPA Method TO-1. Its use in indoor air sampling is described below.

For the Tenax GC^{TM} method air is drawn through a sampling tube packed with Tenax GC^{TM} . The tube may be either glass or stainless steel. The size of the tube can be variable; however, tubes which hold 1 to 1.4 grams of Tenax GC^{TM} (approximately 1.5 cm × 6 cm) are typical. The sampling flow rates are determined by the compounds of interest and the desired sampling time; they can vary from 0.01 liter/min to 1 liter/min. Volatile compounds are thermally desorbed with helium and condensed in a liquid nitrogen cold trap. The sample is then introduced into a gas chromatograph/mass spectrometer for compound identification and quantification.

A key to the successful use of Tenax GC^{TM} is careful cleaning of the sorbent prior to field use to remove background contamination. This can be done in two ways. One way is in the same manner that the tubes are analyzed; i.e., the tube is placed in the thermal desorption unit of the GC/MS system with the carrier gas flowing through the tube. Clean-up times vary, but 48 hours is not unusual. A second technique is to use Soxhlet extraction and vacuum drying at 100 °C. To provide optimum sample integrity, storage and packing of sampling cartridges should be done under clean room conditions, and all samples should be stored in cleaned, sealed containers.

Detection limits for this method are dependent on several factors including the compound, sample volume, background contamination, and instrument detection limits. Sheldon *et al*. (1984) provide a discussion of these factors.

Breakthrough volumes for Tenax GC^{TM} vary according to compound and temperature. Table 4.5 presents data reported by Sheldon *et al*. (1984) and U.S. EPA (1984) on breakthrough volumes for various compounds which can be absorbed by Tenax GC^{TM} . This table shows the great variation across compounds and emphasizes how carefully the sampling volume must be selected to optimize the data being collected. Laboratory and field tests of Tenax GC^{TM} have shown that the precision of the method is poorest for benzene and best for chlorobenzene.

Table 4.5   Examples of Tenax GC^{TM} breakthrough volumes for selected volatile organic compounds[*]

Compound	Breakthrough volume (liters) at specified sampling temperatures		
	70°F[†]	80°F[†]	100°F[‡]
Benzene	54	38	19
Ethylbenzene	693	487	200
Chloroform	9.1	6.6	
Carbon tetrachloride	20	14	8
1,1,1-Trichloroethane	15	12	6
Chlorobenzene	473	303	
Tetrachlorobenzene	196	144	
Trichloroethylene	50	38	20
O-Dichlorobenzene	1463	1139	150
m-Dichlorobenzene	1291	948	
Bromodichloromethane	45	34	
Bromobenzene			300
Xylene(s)			200
Toluene			9.7
Cumene			440
n-Heptane			20
1-Heptene			40
1,2-Dichloroethane			10
Tetrachloroethylene			80
1,2-Dichloropropane			90
1,3-Dichloropropane			150
Bromoform			100
Ethylene dibromide			60

[*]   Based on 1.4 g Tenax GC^{TM} in a tube 1.5 × 6.0 cm.

[†]   *Source* From L.S.Sheldon et al., in *Indoor Air and Human Health*, R.B.Gammage amd S.V.Kaye (eds). Copyright 1985, Lewis Publishers, Chelsea, MI. Used with permission.

[‡]   *Source*: EPA Method TO-1.

Activated carbon.   The activated carbon technique has been traditionally used to monitor the industrial environment. The National Institute of Occupational Safety and Health (NIOSH) as well as ASTM (see Method D 3686–84) have developed methods for the analysis of several compounds using this sorbent. In general, sampling in industrial environments requires a small air volume as the concentrations are high relative to non-industrial environments.

Activated carbon is non-polar. The process by which volatile compounds are collected on the carbon is referred to as chemisorption. This process efficiently collects low molecular weight compounds and other compounds with boiling points less than 300 °C. The sampling cartridge for the activated charcoal is a glass tube, approximately 6 mm by 70 mm. These tubes hold approximately 150 mg activated carbon. Larger tubes are also commercially available. A small personal sampling pump is used to draw air through the cartridge. The analysis method involves solvent desorption and subsequent analysis by gas chromatography with either an electron capture or flame ionization detector, depending on the compound of interest. The most common solvent used for desorption is carbon disulfide; however, methanol, acetone, and 2-propanol are also used. The purity of these solvents is important, especially when sampling for compounds in low concentrations. ASTM Method D 3686–84 provides a listing of compounds which can be collected with activated charcoal, recommended sampling rates, desorption solvent and other analytical parameters.

Stainless steel canisters.   This method collects whole air samples in SUMMA^{TM} passivated stainless steel canisters. The interior of the canister is specially treated by the SUMMA^{TM} passivation process, which leaves a surface of pure chrome-nickel oxide. Samples may be collected under pressurized and subatmospheric pressures. Analysis is performed by GC/MS. Many compounds can be analyzed by this method at concentrations of ppb by volume; they are listed in Table 4.6. This discussion briefly summarizes the method; a detailed description of equipment and appropriate procedures is found in U.S. EPA (1988b).

The sampling procedure involves drawing air into an evacuated SUMMA^{TM} canister through a sampling train which includes a particle filter, mass flow controller, electronic timer, and Magnelatch valve. Both the subatmospheric pressure and pressurized techniques use an evacuated canister. The difference is in the sampling train, where the latter technique includes an additional pump to provide positive pressure. A schematic of the sampling apparatus is shown in Figure 4.6.

After the sample is collected, it is taken to the laboratory for analysis. The canister is attached directly to the GC/MS system. Water vapor is removed with a special dryer, and the VOCs are cooled in a cryogenic trap, and then revolatilized prior to injection into the GC/MS. Other types of detectors can be used with the GC besides a MS; these include an electron capture detector, flame ionization detector, or photoionization detector. The choice of the appropriate detector or

Table 4.6   Compounds which can be detected using treated
stainless steel canisters

Freon 12 (Dichlorodifluoromethane)
Methyl chloride (Chloromethane)
Freon 114 (1,2-Dichloro-1,1,2,2-Tetrafluoroethane)
Vinyl chloride (Chloroethylene)
Methyl bromide (Bromomethane)
Ethyl chloride (Chloroethane)
Freon 11 (Trichlorofluoromethane)
Vinylidene chloride (1,1-Dichloroethene)
Dichloromethane (Methylene chloride)
Freon 113 (1,1,2-Trichloro-1,2,2-Trifluoroethane)
1,1-Dichloroethane (Ethylidene chloride)
cis-1,2-Dichloroethylene
Chloroform (Trichloromethane)
1,2-Dichloroethane (Ethylene dichloride)
Methyl chloroform (1,1,1-Trichloroethane)
Benzene (Cyclohexatriene)
Carbon tetrachloride (Tetrachloromethane)
1,2-Dichloropropane (Propylene dichloride)
Trichloroethylene (Trichloroethene)
cis-1,3-Dichloropropene (cis-1,3-Dichloropropylene)
1,1,2-Trichloroethane (Vinyl trichloride)
Toluene (Methyl benzene)
1,2-Dibromomethane (Ethylene dibromide)
Tetrachloroethylene (Perchloroethylene)
Chlorobenzene (Phenyl chloride)
Ethylbenzene
m-Xylene (1,3-Dimethylbenzene)
p-Xylene (1,4-Dimethylbenzene)
Styrene (Vinyl benzene)
1,1,2,2-Tetrachloroethane
o-Xylene (1,2-Dimethylbenzene)
4-Ethyltoluene
1,3,5-Trimethylbenzene (Mesitylene)
1,2,4-Trimethylbenzene (Pseudocumene)
m-Dichlorobenzene (1,3-Dichlorobenzene)
Benzyl chloride (α-Chlorotoluene)
o-Dichlorobenzene (1,2-Dichlorobenzene)
p-Dichlorobenzene (1,4-Dichlorobenzene)
1,2,4-Trichlorobenzene
Hexachlorobutadiene (1,1,2,3,4,4-Hexachloro-1,3,-Butadiene)

Source: U.S. EPA, Compendium for Determination of Air Pollutants
in Indoor Air, Final Draft, U.S. Environmental Protection Agency,
Research Triangle Park, NC, 1988.

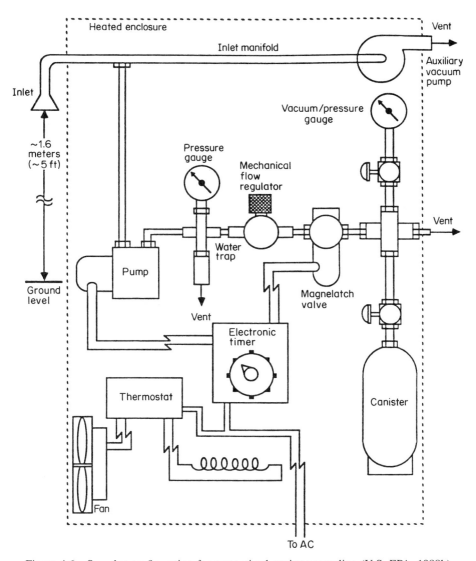

Figure 4.6   Sampler configuration for pressurized canister sampling (U.S. EPA, 1988b)

combination of detectors depends on the specificity and sensitivity of the analysis and the intended purpose of the research.

The advantage of this method is its sensitivity to a wide array of VOCs. The sampling package can be programmed to collect discrete air volumes over an extended time period (e.g., 24 hours) to provide a long term integrated sample. The disadvantage is that the sampling apparatus is large and sophisticated, relative to the solid sorbent techniques. Special care must be taken in the handling and cleaning of the sampling apparatus to avoid contamination of the canister.

Passive Sampling Techniques

Passive samplers for VOCs use solid sorbents and operate on the principle of molecular diffusion. Both activated charcoal and Tenax GC^{TM} have been used in passive sampler badges. In general, exposure times are fairly long, e.g., 30 days.

The activated charcoal badge has been used more frequently than the recently developed Tenax GC^{TM} badge. The passive badge technique has been used by Shields and Weschler (1987) in monitoring organics in office environments, and they have published a method for GC/MS analysis. Lewis and Mulik (1989) reviewed recent developments in passive sampling for VOC and other pollutants and compared results from several different methods and sampling conditions. Several different manufacturers make activated carbon badges including DuPont, 3M, and Abcor. The DuPont badges had the lowest background contamination based on the work by Coutant and Scott (1982).

Monsanto Research Corporation has developed a passive badge which uses Tenax GC^{TM}. The major advantage of the Tenax GC^{TM} is lower sensitivities and the analytical ease of using thermal desorption. However, the Tenax GC^{TM} badge has not been widely used to date, and information on the adsorption capacity and optimum sampling times has not been published.

Portable Gas Chromatographs

Portable gas chromatographs provide real-time data collection in the field. They can serve as screening devices to guide the collection of further samples by other techniques, or when equipped with accessories such as sample loops and data loggers, effectively collect continuous data. The prime disadvantage to this method is generally high limits of detection because there is no preconcentration of the sample as occurs when solid sorbents are used. These limits are usually higher than concentrations found in indoor environments. The two most common portable GCs are made by HNu and Photovac. They both use photoionization detectors. Given the advantage of real-time monitoring, portable GCs may be modified to be of use in indoor, non-industrial environments.

In the next section a selected number of studies on VOCs in the indoor environment are reviewed.

4.9.3 Indoor Concentrations of Volatile Organic Compounds

Numerous studies have been conducted to measure VOCs in various micro-environments. Because of the large class of compounds included as VOCs, it is not useful to attempt the individual review of even a majority of the available literature. The purpose here is to review those studies that best represent research being conducted in this area and then to draw general conclusions regarding research needs. The data from the studies reviewed are presented in Appendix I. Descriptions of these studies are presented by the U.S. EPA (1988a).

Nine studies were selected for review in Appendix I, based on representativeness of the type and number of the microenvironments monitored and reasonable confidence in the sampling and analytical method. Most of these studies were conducted in homes, with some done in schools and office buildings. Although VOC exposure occurs in other microenvironments, e.g., gas stations, vehicles, and dry cleaners, these studies were generally too limited to be considered representative. The sampling technique for the studies discussed below used solid sorbents as trapping media (Porapak Q^{TM}, Tenax GC^{TM}, or activated charcoal); analyses were done using gas chromatography with flame ionization, electron capture, or mass spectrophotometer detectors.

The largest study designed to estimate total exposure to VOCs is EPA's Total Exposure Assessment Methodology (TEAM) study. The method of sampling was with personal samplers using Tenax GC^{TM} followed by GC/MS analysis. Primary goals were to: (1) develop methods to measure individual total exposure to VOC in air, water, and food and calculate the associated body burdens, and (2) use these analytical methods on various representative subjects drawn from several different cities. Although many publications have been generated from these data, only two are presented in Appendix I, Wallace et al. (1985) and Wallace (1986). The sampling and analytical methods were the same for both studies (Tenax GC^{TM} and gas chromatography/mass spectrometry). As part of the former study, homes in New Jersey were sampled. As part of the latter study, individuals in Los Angeles, CA wore personal monitors. Both studies collected ambient data.

The studies summarized in Appendix I are in three tables that present the results in different forms. Because of the extremely large numbers of candidate VOCs and the different groupings measured by different investigators, it was necessary to present the results in these three formats. Table I, Part 1 covers the design parameters for the studies, and Table I, Part 2 presents the results of compounds reported in two or more studies. To facilitate comparisons of results between studies, the data in Table I, Part 2 were further divided into structural categories: aromatic, aliphatic, chlorinated oxygenated and terpenes. Table I, Part 3 presents the results for compounds measured only once among the studies considered. Only three of the nine studies have data in this category. From Tables I, Part 2 and I, Part 3 it is evident that of the numerous VOCs reported (up to 45 in one study) there are relatively few compounds (6) which are reported in most of the studies. Mean and median values (where available) are reported in the results tables. It is recognized that given the variability of VOC concentrations, often over an order of magnitude, means do not best represent central tendency. However, with the number of compounds to review, simplification of the data was mandatory. Further, comparing different measures, such as means, medians, and geometric means, is also not optimum. However, it was found necessary to make comparisons here but with caution, because those were the only values available in the original literature. It is only by critically reviewing results across studies that future research studies can be directed at filling the deficiencies which currently exist. Unfortunately, most studies neither identify specific sources within a microenvironment nor indicate if certain microenvironments have higher concentrations than others. This problem is due primarily to the exploratory nature of most of this research.

The principal findings from the studies listed in Appendix I are:

- A common finding across all the studies in both the U.S. and Europe is that indoor VOC concentrations are higher and consist of more complex mixtures of compounds than are found outdoors.
- VOCs commonly found indoors in concentrations higher than those outdoors are aliphatic, aromatic, and chlorinated hydrocarbons. Indoor concentrations, although higher than those outdoors seldom exceed a fraction of a ppm. For example, a typical value for toluene would be in the range of 20 to 100 $\mu g/m^3$ (5.2 to 26 ppb).
- VOCs inside newer houses are higher than those inside older homes.
- Occupancy levels, indoor activities and indoor furnishings tend to produce elevated indoor VOC concentrations. Smoking tends to be associated with increased indoor concentrations of aliphatic and aromatic VOCs.

4.10 Environmental Tobacco Smoke

4.10.1 Sources and Characteristics

Environmental tobacco smoke (ETS) is the most complex pollutant to which people are frequently exposed. More than 3800 compounds are emitted in both particulate and vapor phases (National Academy of Sciences, 1986a). Concentrations of the chemical and physical constituents have not been characterized fully because individual compounds or groups of compounds depend on numerous parameters, including: the generation rate of the contaminant(s) from the tobacco, the source consumption rate, the air exchange rate, the background concentration of indicator component(s), removal mechanisms (e.g., air cleaning devices, reactions with surfaces, or chemical transformations), dilution volume of the space, and the degree to which the air is mixed.

Total exposure to ETS depends on ETS concentration and the length of time spent in the numerous microenvironments where smoking occurs: homes, offices, public buildings, and various means of transportation (buses, trains, and planes). Exposure to ETS has been measured by various tracers: acrolein, aromatic hydrocarbons, CO, nicotine, oxides of nitrogen, nitrosamines, and inhalable particles.

The most recent and thorough review of ETS exposure is by the National Academy of Sciences (NAS, 1986a). Earlier, the U.S. Surgeon General (1986) reviewed the literature on the numerous tracers used to measure exposure to ETS. These two reviews form the basis of this discussion.

Researchers usually monitor one compound in ETS because it is neither practical nor possible to monitor the full range of compounds. The National Academy of Sciences notes that optimum tracers for ETS should have the following characteristics:

• Unique (or nearly so) to ETS, so that there is minimal contribution from other sources,
• Detectable at low concentrations,
• Similar emission rates among various tobacco products, and
• Consistent ratio between the individual contaminant of interests and the composite pollutant, ETS, under a range of environmental conditions.

No single measure of ETS meets all of these criteria, nor does the scientific community accept one measure as representing ETS exposure. The scientific community also does not agree on the specific components of ETS which may be responsible for the adverse health effects attributed to ETS.

The NAS report does not accept as markers aromatic hydrocarbons, nitrosamines, and oxides of nitrogen. The aromatic hydrocarbons which have been measured in ETS include benzene, toluene, benzo(a)pyrene, and benzo(ghi)perylene. These compounds have numerous sources, both indoors and outdoors, not related to tobacco smoke. Furthermore, their emission rates have not been established. The NAS report also discards acrolein, acetone, and polonium-210 as ETS markers because of low concentrations, existence of other sources, or instability.

Nicotine, carbon monoxide (CO), and respirable particulate matter (RPM) are more acceptable tracers for ETS, but they too have deficiencies. CO and RPM have other indoor sources. CO is not strongly emitted by cigarettes, and is significantly elevated above background only at high smoking rates. However, nicotine is unique to ETS, and its measurement technique has been recently improved (Hammond *et al.*, 1987). Even under conditions of low smoking rates, easily measurable increases in RPM have been recorded above background levels (Repace and Lowery, 1980, 1982).

4.10.2 Available Measurement Methods

Nicotine is an ideal candidate as a tracer for ETS because it is unique to tobacco smoke. However, there are several well-known problems with using nicotine: (1) it is highly reactive, making both sampling and analysis difficult, (2) nicotine can be present in both the particulate and the vapor phases, and (3) particulate or vapor phase nicotine can be re-emitted from surfaces on which it has deposited. Although nicotine is a marker for ETS exposure, the relationship between nicotine and other components of ETS which may be of health consequence has not been established.

This latter problem affects the extrapolation of potential health effects of ETS based on nicotine exposure.

Two approaches are described below for the determination of nicotine in indoor air, one for vapor phase and one for total nicotine. The methods which measure only vapor phase nicotine use a solid sorbent trap. Total nicotine is measured using a treated filter media.

Two different solid sorbents have been used to collect vapor phase nicotine: Tenax GC^{TM} and XAD-2. The Tenax GC^{TM} method (Thompson et al., 1989) uses a packed tube of 200 mg of Tenax GC^{TM} and personal sampling pump capable of drawing 1.7 liter/min. Both the sampling tube and the Tenax GC^{TM} must be carefully cleaned and conditioned prior to use. Nicotine is thermally desorbed from the Tenax GC^{TM} and analyzed by GC with a nitrogen phosphorus detector (NPD). The limit of detection is 0.07 μg/m^3 and limit of quantification is 0.17 μg/m^3 for this method (Thompson et al., 1989).

Breakthrough volume is defined as the volume of air which can be drawn through a trapping media before the analyte of interest is detected in a backup tube. Breakthrough volumes are a function of sample flow rate and air concentration. Thompson et al. (1989) cite breakthrough volumes of 20–45 liters at concentrations of 70–250 μg/m^3. At the sample flow rates used for this method, these volumes are equivalent to sampling times of 11.8 min and 26.5 min, respectively. These test concentrations of nicotine are much higher than those found in typical indoor environments. Breakthrough volumes should be determined at the concentrations which are anticipated in a particular study. Determining breakthrough is done most easily by using a backup section of Tenax GC^{TM} which can be analyzed separately and varying the sampling time. Determination of breakthrough volumes will define appropriate sampling times.

A second type of trapping media used to collect nicotine vapor is XAD-2. The NIOSH method for nicotine uses this sorbent, but the sampling and analytical procedures have a limit of detection of 300 μg/m^3. This is unacceptably high for non-occupational environments. ASTM Committee D22.05 is developing a method which uses XAD-2, and is able to achieve a much lower limit of detection.

The proposed ASTM method uses a 7 cm glass tube containing two sections of XAD-2 separated by a glass wool plug. The front section contains 80 mg and the backup section contains 40 mg of XAD-2. The recommended sample flow rate is 1 liter/min. Nicotine is desorbed from the XAD-2 using a solvent of ethyl acetate and triethylamine. Analysis is by GC/NPD. The limits of detection and quantification are 0.17 μg/m^3 and 1.7 μg/m^3, respectively, for a 1-hour sample. An 8-hour sample can reduce these values by an order of magnitude.

The capacity of the XAD-2 tube is approximately 300 ug of nicotine. Assuming a 1 liter/min sampling rate and an 8 hour duration, the capacity of the tube would be exceeded at concentrations of approximately 625 μg/m^3 (for an 8 hour period).

Total nicotine (vapor and particulate phases) can be collected using a treated filter method as reported by Hammond et al. (1987). Two filters, 37 mm Teflon coated glass fiber, are assembled in series in a personal sampling cassette. The first filter collects total or size fractionated particles. The second filter is pretreated with sodium bisulfate and collects vapor phase nicotine. This configuration collects nicotine which is present in the vapor form in the air and nicotine which can volatilize from particles collected on the front filter. Samples are collected using a personal sampling pump at a flow rate of 1.7 liter/min. An advantage of this configuration is that respirable particle and nicotine concentrations are measured simultaneously with the same sampling apparatus.

A solvent extraction process is used to extract the nicotine from the filters and analysis performed by GC/NPD. Two different extraction procedures are used for the front and back filters. The front filter which contains the particulate matter is ultrasonically extracted with dichloromethane. As this solvent cannot be used with an NPD, a second extraction with heptane is done. Nicotine on the bisulfate treated filter is extracted with a series of solvents: ethanol, sodium hydroxide, and heptane. The treated filter collects nicotine efficiently; a breakthrough analysis in a chamber study indicated less than 1% of the nicotine on the first treated filter was present on the backup filter. Hammond *et al.* (1987) report limits of detection for a sampling flow rate of 1.7 liter/min of 0.2 $\mu g/m^3$ for an 8 hour sample and 2 $\mu g/m^3$ for a 1-hour sample.

Hammond and Leaderer (1987) have modified the sodium bisulfate treated filter method described above from an active sampling technique to a passive one. The method is for vapor phase nicotine and is based on the diffusion of nicotine to a filter treated with sodium bisulfate. Approximately 90% of nicotine in aged ETS is in the vapor phase. The sampling filter cassette consists of a Nucleopore windscreen, sodium bisulfate treated filter and support pad. The filter media is a 37 mm Teflon-coated glass fiber. In a test chamber study, the passive and active sampling methods were compared. The lowest concentration measured accurately was 16 $\mu g/m^3$ over a 5-hour period. Samples were also collected over a one-week period and under variable concentrations. There appeared to be a good correlation between the active and passive techniques which indicated that the nicotine was not off-gassing from the treated filter. The analytical technique for this method is the same as that described for the active method.

An analytical issue common to all of the above methods is that nicotine will adsorb to the glassware used in the collection and extraction procedures. Treating the glassware and the extraction solutions with ammonia suppresses the adsorption of nicotine to glass walls.

Unfortunately, there appears to be little information available on the inter-comparison of the precision and accuracy among the various methods.

4.10.3 Indoor Concentrations

Appendix J summarizes studies on nicotine. Questions regarding the removal rate from the airborne phase and subsequent volatilization back into the air (Rylander,

1985) have limited the comparison of nicotine studies. In addition, the early methods most likely lost nicotine during sampling or analysis, and showed nicotine losses from filters of 80% at sampling flow rates of 4 liter/min (Hammond *et al.*, 1987). This fact may account for the relatively low concentrations reported by Weber and Fischer (1980) as they collected samples at 17 liter/min. Based on studies conducted since 1984, nicotine concentrations of approximately 40 $\mu g/m^3$ represent the upper end of the distribution for such microenvironments as restaurants. Office environments appear to have concentrations less than one-half that value.

4.11 Pesticides

4.11.1 Sources and Characteristics of Pesticides

As a class of indoor pollutants, pesticides have been studied only since the late 1970s. Pesticides are commonly used to control insect pests such as termites, Japanese beetles and fungi in both the indoor and outdoor environments. Exposure to pesticides can occur through a variety of routes: inhalation, dermal absorption, and ingestion. Although there are occupational standards for worker exposure to a variety of pesticides, there are no standards for the non-occupational environment. In 1982, the National Academy of Sciences Committee on Toxicology reviewed the health effects of four termiticides (NAS, 1982). They concluded that there were insufficient data to assess health risks but recommended interim guideline concentrations of 5, 2, 1, and 10 $\mu g/m^3$ for chlordane, heptachlor, aldrin/dieldrin, and chlorpyrifos, respectively.

Pesticides can be grouped into three broad categories based on application procedure: subterranean, airborne, and impregnation (e.g., wood preservation). Pesticide exposures most frequently occur in residences, particularly of slab construction, although exposures are not limited to that type of structure. Airborne concentrations depend on a host of factors including vapor pressure, application technique, and time since application. One general observation which can be drawn from a review of the current literature is that in geographical areas where whole house application of pesticides is routine, further study is required to determine the range of possible exposure.

4.11.2 Available Measurement Methods for Pesticides

Many different techniques for pesticide sampling have been published in the scientific literature. Various collection media have been used with solvent extraction and analysis by gas chromatography (GC) with various detectors. The types of solid sorbents used include various GC column packing materials, polyurethane foam (PUF), and charcoal. Unfortunately, little information is available on a cross comparison among the various sorbents used; thus it is difficult to evaluate the relative precision and accuracy of the different techniques.

Table 4.7 summarizes numerous collection and analytical methods which have been used to measure various pesticides. The methods are grouped into three main categories based on sampling method: solid sorbents, mixed media, and bubblers. Historically, the most frequently used media was Chromosorb 102. However, the U.S. EPA has put considerable resources into developing the PUF sampling technique. It has also undergone field application through its use in the Non-Occupational Exposure to Pesticide Study (NOPES) (Lewis *et al.*, 1986). PUF also appears to collect the broadest spectrum of pesticides. These are fully listed in the cited article as well as in the ASTM Method of Committee D-22.05. A trapping media which has the ability to collect a broad spectrum of pesticides is extremely useful for survey studies.

Selection of an appropriate sampling media depends on several issues. These include: compound(s) of interest, ease of use in the field, analytical preparation of the media, and detection limits. An advantage of the methods which use GC packings is that standard size glass tubes can be used to make the trap; for the PUF technique special glass holders must be fabricated.

The methods used to analyze pesticides vary according to media and pesticide. In general, the methods involve separation by gas chromatography and various detectors, ranging from electron capture detectors to mass spectrometers. The mass spectrometer detectors have the potential for providing positive identification of specific compounds. A possible analytical problem with these methods (particularly PUF) is the contamination of glassware and sampling apparatus with traces of PCBs or pesticides. This is especially important if low concentrations are expected. The ASTM method presents several techniques for minimizing such interferences. In addition to avoiding contamination in the sampling apparatus, sampling media (e.g., PUF) must be rigorously cleaned (two Soxhlet extractions for 14 to 24 hours, each) prior to use. The ASTM method notes that Tenax GC^{TM} can be sandwiched between two PUF plugs to collect compounds with saturation vapor pressures greater than 10^{-3} mm Hg.

4.11.3 Indoor Concentrations

Numerous studies on pesticide concentrations in different microenvironments have been conducted. Appendix K summarizes the results of these studies. Certain microenvironments present elevated exposures, especially those associated with the pesticide application business. Another important parameter which influences exposure is the rate at which the compound decays. In a study by Wright *et al.* (1981) chlorpyrifos and diazinon had the slowest decay rates, measured after four days.

Exposure to the termiticide chlordane has been measured in many houses and apartments located on military bases. In the initial survey of Livingston and Jones

Table 4.7   Pesticide sampling methods

Sampling method	Analytical method	Compound(s) reported	Reference
• Solid Media GC Dura Pak-Carbo Wax 400/Porasil F	Diethyl ether extraction and GC/ECD	Ronnel Lindane Carbaryl Diazinon	Melcher et al. (1978)
Chromosorb 102	Hexane-acetone extraction and GC/ECD	11 pesticides (including organochlorine and organophosphorous compounds)	Thomas and Seiber (1974)
Polyurethane foam (PUF)	Diethylether in hexane GC with various detectors including: ECD, NPD, FPD, HECD, and MS. Also HPLC with UV detector.	Collection efficiencies have been determined for: 17 organochlorine pesticides 3 PCB mixtures 28 organophosphorous, organo-nitrogen, and pyrethroid pesticides	Lewis and MacLeod (1982)
Polyurethane foam (PUF)	GC/FPD	Acephate	Wright et al. (1981)
Charcoal	CS_2 extraction and GC/ECD	Lindane	Seifert and Schmahl (1987)
• Mixed media PUF and glass fiber filters	Filters: liquid CO_2 extraction PUF: Freon 12, separation by HPLC and GC/MS analysis	Lindane Pentachlorophenol	Oehme and Knöppel (1987)
Porapak R+ glass fiber filters	Benzene or ethylacetate extraction (depending on analyte) and GC/FPD	Pest control strips: Diazinon Propoxur Chlorpyrifos	Jackson and Lewis (1981)
• Bubblers Hexylene glycol	HPLC/UV (Solvent: 2 propanol)	Bendiocarb Carbaryl Propoxur	Wright et al. (1981)
	GC/FPD	Diazinon Fenitiothion Chlorpyrifos	
Potash solution	GC/ECD	Pentachlorophenol	Hümpel and Keller (1987)

GC, Gas chromatography; ECD, Election capture detector; NPD, Nitrogen phosphorous detector; FPD, Flame photometric detector; HECD, Hall

(1981) 60% of the units had concentrations greater than 1 $\mu g/m^3$. The heating ducts were located in the slab or immediately adjacent to it. Hypothesized chlordane migration pathways included: direct contamination by the termiticide in the heating duct or diffusion through cracks or joints in the ducts.

Wood preservatives, lindane and pentachlorophenol, also present indoor exposures. Although only limited data exist, research has shown that these compounds can migrate into other household materials (e.g., textile materials) from the treated wood (Gebefugi and Korte, 1984). This implies that wood preservatives can present a long-term exposure.

The Non-Occupational Exposure to Pesticides Study is the first study of pesticide exposure in randomly selected homes. In the results of the pilot phase of this study (Lewis *et al*., 1986), the four most common pesticides identified were chlorpyrifos, diazinon, chlordane, propoxur, and heptachlor. These data indicate that general pesticide usage in homes may represent a source of exposure.

In summary, the following conclusions from a review of the studies of various pesticides and microenvironments can be drawn:

• Subterranean application of pesticides produces detectable concentrations of pesticides indoors.
• Preliminary results from NOPES indicate that general pesticide usage in homes may present exposures.
• Initial data indicate that wood preservatives (lindane and PCP) may present a source of exposure long after application.

4.12 Odors

4.12.1 Sources and Characteristics of Odors

Odors are a unique class of indoor contaminants since their detection and measurement are based on the human olfactory system rather than some type of instrument using physical or chemical principles. Specific odorous materials such as hydrogen sulfide are amenable to direct chemical measurement, but most odor problems, whether indoors or outdoors, are commonly caused by complex mixtures of organic compounds of which one or more individual compounds have low odor thresholds. Odors have both outdoor and indoor sources. Outdoor odor problems that may come indoors include odors from rendering plants or sewage treatment plants. Indoor-generated odor problems include body or tobacco smoke odors, or odors from the out-gassing of building materials or furnishings.

The following paragraphs are only a brief summary of the extensive writings on the fascinating subject of odor perception and measurement. Those readers interested in pursuing the subject further are directed to ASHRAE (1985), Moncrieff (1967), Cain (1976), and Hooper and Cha (1988) and the reference lists in these publications. This chapter will not cover other types of sensory effects such as various types of irritation (eye, nose, throat, and skin).

The sense of smell is an important part of our lives. The smell of an appetizing meal being prepared enhances the enjoyment of good food and a pleasing perfume heightens the attraction of one person to another. Malodors, on the other hand, warn us of unsafe or unsavory conditions such as spoiled food or an overcrowded office with poor ventilation. In the outdoor environment one usually has freedom to move away from disagreeable odorous situations, or changing wind direction may carry odors away. However, in an indoor setting, a person may be confined for prolonged periods (e.g., 8-hour work days, 5 days a week) in a room or an area of a building in which disagreeable odors are present. This perception of "entrapment" in an indoor space has profound physiological and psychological effects on those who must endure poor indoor air quality. Thus, the concentrations of odorous and irritating materials that cause dissatisfaction tend to be much less indoors than outdoors.

The perception of odor has four major dimensions or descriptors: detectability, intensity, character, and hedonic (pleasant or unpleasant) tone (Hooper and Cha, 1988).

Odor Threshold

The threshold of detectability of an odor (odor threshold) is the theoretical minimum concentration of odorant stimulus necessary for perception in some specified percentage of a population exposed to the odorant. Odor panels consisting of a number of persons with "normal" ability to detect odors are used for making odor threshold measurements. Relative concentrations of odorants are determined by the volume of odor-free air required to dilute the odorous air sample to the odor threshold. Typically, the odor threshold is reached when 50% of the panel members are just able to detect the odor. Thus, odor thresholds are not fixed physical constants but are statistical points representing the best estimate value from a group of individual scores. Hooper and Cha (1988) discuss odor panel selection and the optimum number of panel members.

Odor thresholds for pure compounds are presented in a number of references, but for any one compound these published threshold values may vary over several orders of magnitude depending largely on the methodologies used for determining the thresholds (Verschueren, 1983). The American Industrial Hygiene As -sociation reviewed data on published odor thresholds for compounds for which Threshold Limit Values (TLVs) have been published (AIHA, 1989). They utilized data developed from studies which met well-defined rigorous standards for the methodologies used such as number of panelists, methods of sample presentation, etc. The results of this work have greatly narrowed uncertainty ranges in odor thresholds for the materials covered. Nevertheless, published odor thresholds should be used with caution and only after the methods for their determination have been carefully scrutinized.

Odor Intensity

The intensity of an odor above its threshold increases with the concentration of the odorant in accordance with the following psychophysical power function:

$$I = kC^n$$

where:

I = perceived odor intensity
k = y-intercept of the psychophysical function
C = concentration of odorant
n = slope of psychophysical function

The exponent of this power function is a critical parameter in olfactory science since it varies from one odorous compound to another and is an indicator of the relative effectiveness of dilution for the control of indoor or outdoor ambient odors. The exponent is less than 1.0 and varies roughly between 0.2 and 0.7, but within this range the relative amount of dilution air required to dilute an odorous material to its odor threshold varies tremendously. For example, with an exponent of 0.7, to reduce perceived odor intensity by a factor of 5, the concentration must be reduced by a factor of 10. On the other hand, with an exponent of 0.2, a five-fold reduction in odor intensity will require more than a factor of 3000 reduction in concentration (ASHRAE, 1985). This feature of the odor intensity power function has important implications in indoor air quality as it relates to the relative effectiveness of dilution ventilation in solving odor problems.

Various methods are used to assess the relative strength of an odor. The most usual method is to grade an odor in accordance with a scale such as that presented by ASHRAE (1985):

0 = no odor = threshold or just recognizable
1 = slight odor
2 = moderate odor
3 = strong odor

Another method is to compare odor intensity of an environmental odor with a standard odor such as 1-butanol by matching the intensity of a 1-butanol odor with that for the environmental odor. Figure 4.7 is a graph of an odor intensity function for 1-butanol, an odorant commonly used as an odor standard. This method is described by the American Society for Testing Materials (ASTM) as ASTM E544–75, *Standard Recommended Practices for Referencing Suprathreshold Odor Intensity*.

Perhaps the most widely used method for scaling odor strength is to describe the

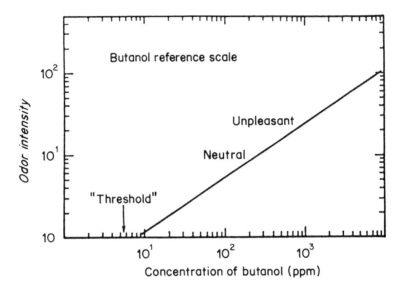

Figure 4.7   Standardized function relating perceived magnitude to concentration of 1-butanol. (Reproduced from ASHRAE (1985) with permission of the American Society for Heating, Refrigerating and Air Conditioning Engineers)

odor in terms of odor units. This is a measure of the number of volumes of odor-free air required to dilute a given volume of odorous air to the odor threshold, and is commonly used for determining the degree of odor control that may be required for a given odor source. However, it must be recalled that the relative number of odor units to scale an odor do not necessarily bear any relationship to perceived odor magnitude because of the wide variability of values for the exponent in the above power function for various odorants.

Odor Character and Acceptability

There are various lists of characteristics for qualitatively describing odors. Hooper and Cha (1988) describe them in detail, and they include such terms as fishy, musty, sweaty, rancid, sewer, etc. The acceptability or hedonic tone of an odor is also a subjective factor. However, the acceptability of an odor, particularly in the indoor setting, is often irrelevant. Perception of odor is based on a combination of frequency of occurrence, odor character, and odor intensity. Even very agreeable odors such as perfume or roasting coffee may become objectionable if exposure to them persists. In addition, prolonged exposure to certain odors causes a reduced ability to perceive odor or "odor fatigue." An example of such an odorant is H_2S. This property is especially critical in view of the high toxicity of H_2S.

4.12.2 Available Methods for Measuring Odors

As pointed out earlier, most environmental odor problems are caused by complex mixtures of organic gases and vapors. This is especially true indoors where occupants of buildings are exposed to odors generated indoors from such sources as smoking, human bioeffluents, indoor furnishings and construction materials, cooking emissions, and a wide variety of consumer products. While it is possible to collect the organic compounds that create these odors by various concentrating methods such as adsorption or cryogenics (see Section 4.9 on Volatile Organic Compounds), the analysis for these materials is costly and time-consuming, and the results are often difficult to interpret because of the large number of materials involved and the wide range of odor thresholds. Therefore, odor measurement generally involves the collection of representative air samples and then subjecting them to sensory analysis by a trained odor panel. Normally the odor strength is measured in terms of dilution-to-threshold (D/T) to determine the number of odor units for the samples. Such measurements are made with forced-choice triangle olfactometers. In these systems an odor panelist is presented with a series of samples to test. Each test consists of three samples, two of which are activated carbon filtered pure air and the third is a sample of the odor-containing air at some known dilution. Therefore, the panelist is forced to make a choice of which of the three samples has a detectable odor (Dravnieks, 1975). Figure 4.8 is a diagram of a portable version of this system used by TRC Environmental Consultants, Inc. for field studies (Hooper and Cha, 1988). In using these systems, samples of increasing odorant concentration are presented to the odor panel. If suprathreshold samples are introduced first, the odor threshold is likely to be obscured. The odor threshold and D/T point are based on a log-log curve of dilution factor and odorant concentration, if known, versus percent of panel detecting the odor at the point where 50% of the panel just detect an odor. Figure 4.9 is a typical odor response curve for an odor panel for two samples containing 500 ppm of 1-butanol. The American Society for Testing Materials has developed ASTM E679–79, *Standard Practice for Determinations of Odors and Taste Thresholds by a Forced-Choice Ascending Concentration Series Method of Limits*.

Berglund *et al*. (1984) have developed a unique mobile laboratory for conducting research on odors and other sensory effects in buildings. Figures 4.10 and 4.11 are diagrams of this facility which consists of three basic components contained in standard 20-foot freighter containers: environment chamber, olfactometer, and sampling and HVAC equipment. The environment chamber is of double wall construction using aluminum and stainless steel for air contact surfaces. Tempered air is circulated in the wall interspace to control surface temperature of the inner room and to prevent infiltration of outside ambient air into the inner space. The second chamber contains the olfactometer with a small laboratory for chemical analysis of air pollutants. Known mixtures of sample gas or reference gases (e.g., pyridine) can be made with odor free air for determination of odor thresholds

118

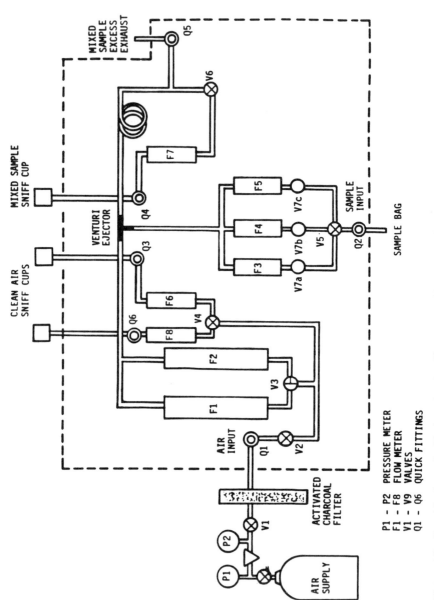

Figure 4.8 Portable dynamic dilution triangle olfactometer (Hooper and Cha, 1988)

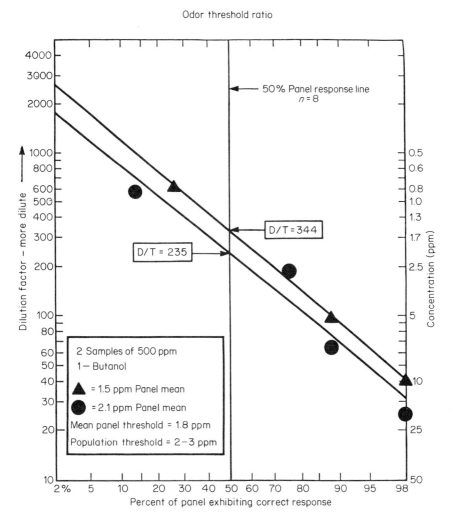

Figure 4.9 Typical odor threshold response curves for butanol dilutions (Hooper and Cha, 1988)

using human subjects. The third chamber contains the three HVAC systems serving the environment chamber. One HVAC system provides cleaned air from which odorants have been removed, another HVAC system is used for air sampling from a building, and the third HVAC system provides air for wall temperature control in the environment chamber.

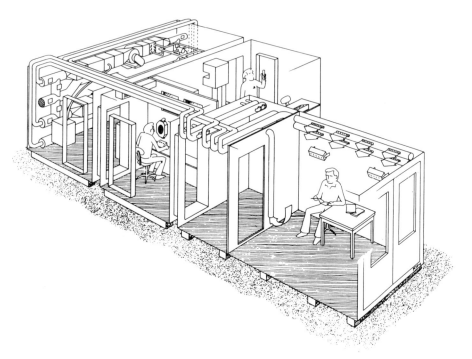

Figure 4.10   Three-unit mobile environment laboratory consisting of environment chamber, olfactometer and sampling system with HVAC equipment. (Reproduced from Berglund *et al*. (1984) with permission of Pergamon Press, Inc.)

4.12.3 Indoor Measurement of Odors

In spite of the importance of odors in judging the acceptability of indoor air quality, few major studies of indoor concentrations of specific pollutants have included odor measurements.

Duffee *et al*. (1980) conducted a study on the effect of reduced ventilation on indoor perception of odor in a number of buildings (schools, hospitals, and office buildings) located in Connecticut and California, U.S.A. Odor panels were used for quantifying odor levels, and parallel samples were collected on Tenax GC™ absorbent for analysis of the VOC content of the building's air. The buildings were operated under both "normal" and "reduced" ventilation conditions. The "reduced" condition was usually less than half of the air exchange rate for the "normal" condition. In spite of significant increases in measured VOC concentrations under "reduced" ventilation conditions, the authors showed that in the absence of smoking in the buildings, it is possible to reduce ventilation rates significantly without producing unacceptable odor levels. However, there were inconsistencies in the comparative perceptions of visitors to the buildings and occupants.

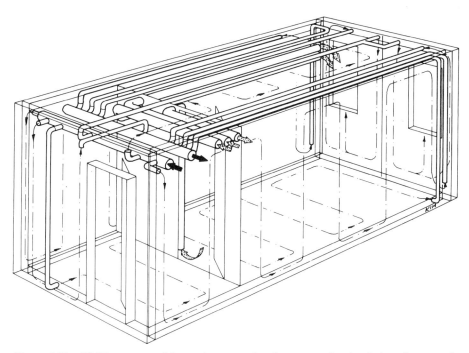

Figure 4.11   Wall interspace of the environment chamber system for circulating of tempered
air for the control of interspace pollutants and surface temperature of the inner room.
(Reproduced from Berglund *et al*. (1984) with permission of Pergamon Press, Inc.)

Minamino (1987) studied odors in several university buildings and compared
the perception of odors with CO_2 concentrations and found that there was a
good correlation between odor sensation and CO_2 concentration. However, students
(occupants) were not able to discriminate between the types of odors even at CO_2
concentrations as high as 5000 ppm. Odor panelists (visitors), on the other hand,
were able to distinguish the types of odors, principally body odors.

Fanger (1987) has introduced a useful term the "olf" (abbreviation of *olf* action)
as a measure of indoor pollution sources that impact the perception of indoor air
quality. The olf is "the emission rate of bioeffluents from a standard sedentary
person in thermal comfort." This term recognizes the importance of body odor
in occupied, inadequately ventilated buildings in the perception of poor indoor air
quality. The author believes that using this unit can provide more accurate standards
for ventilation than physical measurements such as CO_2 concentration.

Lauridsen *et al*. (1987) in a study of odors in 15 offices and 5 conference rooms
used a team of 49 judges (visitors) to judge indoor odors. Observations of odors
were made under three different conditions: with no mechanical ventilation and
no occupants, with mechanical ventilation and no occupants, and with mechanical

ventilation and occupants. The results of the study showed that background odor is an important characteristic of indoor air quality and may be more important than bioeffluents or tobacco smoke in creating dissatisfaction with indoor air quality. The work also demonstrates that use of physical measurements of ventilation and ventilation effectiveness such as CO_2 concentrations are not necessarily the best basis for establishing acceptable indoor air quality.

4.13 Carbon Dioxide (CO_2)

4.13.1 Sources and Characteristics of CO_2

CO_2 is a normal constituent of the world's atmosphere (see Table 1.1). Its present concentration in remote outdoor locations is approximately 350 ppm. However, worldwide concentrations have been increasing slowly over the years principally because of carbonaceous fuel combustion and also perhaps the reduction of forested areas. Increases in ambient concentrations of CO_2 and other pollutants are creating fears about global warming from the "greenhouse effect." In the center of large cities where large amounts of fuel are burned for power generation, heating, and vehicular propulsion, outdoor ambient CO_2 concentrations under stable atmospheric conditions can easily reach levels twice those found in remote locations. Although somewhat soluble in water, CO_2 is quite unreactive and readily penetrates the indoor environment from outdoors. Thus, in the absence of indoor sources, concentrations indoors follow outdoor concentrations.

Indoor sources of CO_2 are principally indoor combustion sources (e.g., unvented stoves and heaters, smoking, and attached or basement garages) and human respiration. A typical human at rest produces 0.30 liter/min of CO_2. Therefore, a measurement of CO_2 in an occupied space (e.g., office building) is an indication of the adequacy of fresh air ventilation. The ASHRAE Standard for "Ventilation for Acceptable Indoor Air Quality" (ANSI/ASHRAE 62–1989) calls for a ventilation rate of 15 CFM/person in mechanically ventilated buildings (ASHRAE, 1989). While this ventilation rate is designed primarily to control tobacco smoke odor, it corresponds to an indoor CO_2 concentration of approximately 1000 ppm including a background CO_2 concentration of 300 ppm. An indoor concentration of 1000 ppm CO_2 is commonly used as a rough dividing line between adequate and inadequate fresh air ventilation rates.

CO_2 can be tolerated in concentrations up to about 1% (10000 ppm). The 8-hour time-weighted average occupational standard used in the U.S. is 5000 ppm. Extremely high CO_2 concentrations trigger increased breathing rates and ultimately cause narcosis. Such excessive concentrations can occur in confined spaces such as unvented vaults and wine cellars where heavier-than-air CO_2 may accumulate from bacterial actions on organic materials.

CO_2 as a common indoor pollutant in occupied buildings in and of itself, probably has little direct physiological effect on occupants in concentrations normally found. However, elevated concentrations (e.g., >1000 ppm) indicate inadequate ventilation and elevated concentrations of more important pollutants such as VOC.

Because of CO_2's unreactivity, it can often serve as a convenient tracer for measuring air exchange rates as described in Chapter 3.

4.13.2 Available Measurement Methods for CO_2

CO_2 at extremely high concentrations (several percent) can be measured by the Orsat Method commonly used in stack sampling in which CO_2 is absorbed in KOH solution and estimated by volume differences. In an early edition of the NIOSH Manual of Analytical Methods (NIOSH, 1977) a method is presented which is based on bag sampling and subsequent GC analysis using a thermal conductivity detector. The working range of the method is estimated to be 500–1500 ppm. The principal problem with this method aside from its requiring rather involved sampling and analytical procedures is that any material having the same retention time in the GC column as the analyte will interfere. Stain-indicating detector tubes commonly used by industrial hygienists may also be used if concentrations are sufficiently high.

Since CO_2 absorbs strongly in the infrared region, non-dispersive infrared (NDIR) spectrophotometry is an excellent method for its accurate measurement. Several commercial instruments are available based on this principle. Fixed recording NDIR instruments are useful for developing diurnal patterns of CO_2 in buildings, and small portable instruments are especially useful in surveying CO_2 levels in buildings to provide an initial indication of ventilation rates and ventilation effectiveness.

4.13.3 Indoor Measurement of CO_2

Measurements of CO_2 indoors are commonly made in SBS studies or other building ventilation studies. Since CO_2 is not normally of concern as an important pollutant, but rather as an indicator of inadequate ventilation, extensive databases on indoor CO_2 levels similar to databases on indoor pollutants are not readily available. Nevertheless, several studies are worthy of mention since CO_2 measurements feature prominently.

In a major study of the effects of reduced ventilation in a large office building, Turiel et al. (1983) measured indoor air quality in terms of several pollutants during two different ventilation conditions: 100% outside air and 15% outside air. CO_2 measurements were made semi-continuously at four locations in the building using a single NDIR instrument which sampled the four areas sequentially for 10-minute intervals. Indoor CO_2 concentrations varied directly with occupancy and reached maximum values of approximately 800 ppm with 100% outside air and approximately 1600 ppm with reduced ventilation. Some of the contaminants measured ("fine" particulate matter, hydrocarbons and formaldehyde) showed increased indoor concentrations with reduced ventilation of the same order as CO_2, but odors judged by building occupants and odor panelists were essentially the same under both conditions.

Fecker *et al*. (1987) carried out studies on comfort and acceptability of indoor air quality based on body odors in both a climatic chamber and a lecture hall. The parameters measured were CO_2 concentrations and temperature. It was found that the upper limit for CO_2 is 1500 ppm since at this concentration no more than 15% of the occupants complained of unpleasant odor. However, 30 to 40% of people who entered the chamber or lecture hall found odors unacceptable at this CO_2 concentration. The authors warn that CO_2 fails as an indicator of acceptable indoor air quality if there are sources of indoor pollutants and odors other than human occupants.

Konopinski (1984) measured CO_2, formaldehyde and temperatures in a large number of homes. He used detector tubes for CO_2. The 57 indoor CO_2 samples yielded a mean concentration of 734 ppm and a maximum of 3000. In view of some of the high indoor values measured, a further study (Konopinski, 1989) was made in a single residence using a continuously recording portable NDIR CO_2 monitor. Indoor CO_2 concentrations were monitored in several rooms of the house and in relation to several activity patterns. Peak CO_2 in the kitchen-family room were of the order of 2500 during occupied periods. Peaks of this same order were found in the bedroom during sleeping periods. The author inferred from these data that during the night and depending upon the sleeping position human exposures to CO_2 could be much higher than depicted by sampling of general room air.

4.14 Bioaerosols

4.14.1 Sources and Characteristics

Bioaerosols are defined by the American Conference of Governmental Industrial Hygienists (ACGIH, 1989b) as airborne particles, large molecules, or volatile compounds that are living or were released from a living organism. Bioaerosols represent a heterogeneous category of indoor pollutants, comprised of bacteria, endotoxins, fungi, protozoa, and antigens. Bioaerosols also vary in size, ranging from less than 0.1 μm to 100 μm. Figure 4.12 represents the relative sizes of several different types of bioaerosols. Because microbiological aerosols are so diverse, the sampling and analytical techniques used to measure them are also diverse. A good reference on this topic is *Guidelines for the Assessment of Bioaerosols in the Indoor Environment* (ACGIH, 1989b), which forms the basis for the following discussion.

Before describing the various types of bioaerosols and sources of bioaerosols, it is useful to understand the three conditions necessary for bioaerosols to become airborne. First is the presence of a reservoir, such as living hosts for bacteria (e.g., *Legionella*). Second is that the organism can amplify, i.e., increase in number or concentration. This requires that the conditions of the reservoir be conducive to

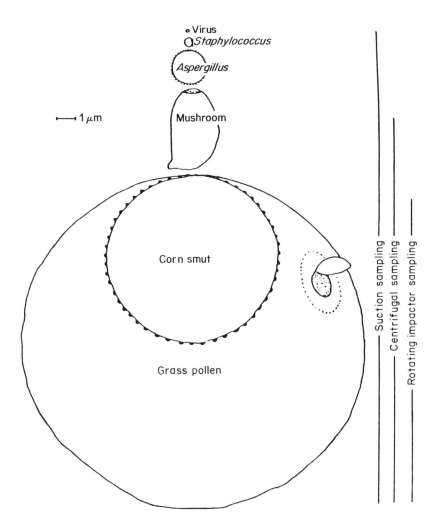

Figure 4.12 Diagrammatic representation of the size range of airborne biological particulates and optimum efficiency ranges for three sampling modalities. (Reproduced from Burge and Solomon (1987) with permission of Pergamon Press, Inc.)

growth. Third is a mechanism for dissemination or aerosolization, e.g., an HVAC system which includes a humidification component.

The five major types of microbiological aerosols of concern for indoor air quality are bacteria, endotoxins, fungi, viruses, and antigens.

Bacteria. Bacteria are prokaryotic (no distinct cell nucleus) organisms that have a cell membrane, DNA, and some subcellular components. They are naturally occurring in both the indoor and outdoor air. Various types of bacteria important to indoor air quality are shown in Table 4.8. Although some bacteria are common in indoor air, it is the disease producing bacteria that are of concern. Indoor conditions can amplify and disseminate them, presenting an inhalation exposure.

Table 4.8   Some bacteria common in indoor air or causing airborne disease

Organism	Gram	Spores	Disease	Occurrence in air Outside	Inside
Legionella pneumonophila	−	No	Legionellosis	Rare	Rare
Pseudomonas	−	No	Pneumonia	Occasional	Rare
Entorics	−	No	Various	Rare	Rare
Bacillus	+	Yes	Hypersensitivity Pneumonitis	Common	Common
Micrococcus	+	No	Usually none	Common	Common
Mycobacterium	(+)	No	Tuberculosis	Common	Common
Thermoactinomyces	+	Yes	Hypersensitivity	Rare	Rare
Corynebacterium	(+)	No	Usually none	Rare	Occasional

Source: Guidelines for the Assessment of Bioaerosols in the Indoor Environment, ACGIH, 1989. (With permission of the American Conference of Governmental Industrial Hygienists.)

Endotoxins.   Endotoxins are contained in the outer membrane of Gram negative bacteria. The Gram-stain is a procedure used to separate bacteria into two groups. Endotoxins are highly toxic and stimulate the immune system of susceptible individuals, causing such diseases as humidifier fever.

Fungi.   Fungi are eukaryotic organisms, containing one or more nuclei and other membraneous organelles. They can be unicellular such as yeasts or multicellular such as molds. Most fungi produce spores which can become airborne. Fungi can be saprophytic (obtain nourishment from non-living matter), parasitic (live off another living organism), or both. Sources of fungi are listed in Table 4.9. Only a few fungi invade living cells and cause infectious disease, usually in people with depressed immune systems. Most fungi produce metabolic products, e.g., proteins, that can induce an antibody response and cause hypersensitivity diseases.

Viruses.   Viruses are intercellular parasites and are the smallest of all life forms. They are replicating organisms, containing RNA or DNA, but not both, and are dependent on the host cell for reproduction. The sources of viruses are the host organism, e.g., humans, not air handling systems. Common viruses are influenza, measles, and chickenpox. Risk of disease depends on several factors, including susceptibility of individual, virus virulence and concentration, and particle size. Air sampling is not usually conducted for viruses, as disease diagnosis is the best indicator of exposure.

Table 4.9   Health effects and sources of a few fungi

Organism	Causal part	disease	Usual source
Histoplasma	Spores	Histoplasmosis	Bird droppings
Cryptococcus	Spores	Cryptococcosis	Bird droppings
Aspergillus	Spores	Invasive aspergillosis; aspergilloma; allergic bronchopulmonary aspergillosis (ABPA)	Self-heated compost; damp organic material
Aspergillus Penicillium Sporobolomyces	Spores	Hypersensitivity pneumonitis (HP)	Damp organic material; standing water
All airborne fungus spores	Spores	Allergic asthma; rhinitis	Damp organic material; outdoor air
Aspergillus flavus	Toxin	Cancer	Agricultural products
Stachybotrys atra	Toxin	Acute toxicosis	Damp cellulosic material
Ulocladium	VOC	Headache	Damp organic material
Penicillium species	VOC	Irritants	Damp organic material

VOC, volatile organic compounds.
Source: Guidelines for the Assessment of Bioaerosols in the Indoor Environment, ACGIH, 1989. (With permission of the American Conference of Governmental Industrial Hygienists.)

Antigens.   Antigens are compounds, usually proteins, produced by living organisms, which induce a detectable immune response from exposed individuals. The types of diseases which antigens cause are hypersensitivity pneumonitis (HP), allergic asthma, allergic rhinitis (runny nose), and allergic aspergillus. Antigens come from a variety of sources (see Table 4.10). In buildings with HVAC systems, the most common sources are humidifiers with reservoirs that support microbial growth. In residences, the common sources are house dust mites, cats, and cockroaches.

Table 4.11 summarizes the common types of bioaerosols, their characteristics, health effects, and indoor sources. It is evident from this table that a common characteristic of the various sources is the presence of moisture. When conducting an indoor air evaluation of a building or a residence, sampling for biological aerosols is appropriate only when there is evidence of a source. Frequently, it is appropriate to sample and analyze both the indoor air and the source material. Air sampling in such instances is most useful in estimating the contribution of the suspected source to the indoor air.

Table 4.10   Sources of some common airborne antigens

Organism	Antigenic product	Common sources
Bacteria	Organisms, soluble antigens	Humidifiers
Fungi	Organisms, spores, fragments, soluble antigens	Outdoor air, indoor surfaces, humidifiers, house dust
Protozoa	Soluble antigens	Humidifiers
Anthropods		
Mites	Fecal particles	House dust
Cockroaches	Fecal particles, body parts	House dust
Mammals	Skin scales, saliva, urine	House dust

Source: *Guidelines for the Assessment of Bioaerosols in the Indoor Environment*, ACGIH, 1989. (With permission of the American Conference of Governmental Industrial Hygienists.)

4.14.2 Available Measurement Methods

There are several different methods which can be used to sample microbial aerosols. However, before the investigator considers collecting air samples, the following admonition from ACGIH should be noted, "One should use air sampling only as a last resort, except in research situations" (ACGIH, 1989a). There are many reasons for this:

- Selection of the appropriate collection media is critical and no one method is appropriate for all organisms.
- Great care must be used in collecting the samples to avoid contamination.
- Analysis of the samples requires a highly skilled laboratory which has experience in identifying environmental microbial aerosols and perhaps environmental mycology. The analyses are expensive.
- Interpreting the results in terms of health effects is difficult as there is minimal dose–response information available.
- Comparing concentrations between different studies must be done with caution because the collection methods may not be comparable. In addition, the data are highly variable.
- Guidelines have not been established which indicate "safe" concentrations of the various bioaerosols.

Notwithstanding the limitations of the data available from monitoring for bioaerosols, in certain situations conducting air sampling is called for. The following discussion highlights the available methods; they are applicable to sampling for bacteria, fungi and some antigens. The same general methods apply to viruses, but sampling for viruses is not usually done because well characterized symptoms indicate the presence of the virus. Methods for sampling endotoxins are

Table 4.11 Characteristics and sources of common bioaerosol components

Living source	Airborne unit	Examples or sources	Primary human effects	Principal indoor sources
Bacteria	Organisms	*Legionella*	Pneumonia	Cooling towers, Hot water sources, hot damp surfaces
	Spores	*Thermoactinomyces*	Hypersensitivity pneumonitis (HP)	Stagnant water reservoirs
	Products	Endotoxin	Fever, chills	
		Proteases	Asthma	Industrial processes
Fungi	Organisms	*Sporobolomyces*	Hypersensitivity pneumonitis (HP)	Damp environmental surfaces
	Spores	*Altamaria*	Asthma, rhinitis	Outdoor air, damp surfaces
	Spores	*Histoplasma*	Systemic infection	Bird droppings
	Antigers	Glycoproteins	Asthma, rhinitis	Outdoor air
	Volatiles	Aldehydes	Headaches, mucous membrane irritants	Damp surfaces
Protozoa	Organisms	*Naegloria*	Infection	Contaminated water reservoirs
	Antigens	Acanthamoeba	Hypersensitivity pneumonitis (HP)	Contaminated water reservoirs
Viruses	Organisms	Influenza	Respiratory infection	Human hosts
Algae	Organisms	*Chlorococcus*	Asthma, rhinitis	Outdoor air
Green plants	Pollen	Ambrosia (ragweed)	Asthma, rhinitis	Outdoor air
Anthropods	Feces	Dermatophagocides (mites)	Asthma, rhinitis	House dust and high humidity
Mammals	Skin scales	Horses	Asthma, rhinitis	Horses
	Saliva	Cats	Asthma, rhinitis	Cats

Source: Guidelines for the Assessment of Bioaerosols in the Indoor Environment, ACGIH, 1989. (With permission of the American Conference of Governmental Industrial Hygienists.)

at the research stage and are not discussed further. The techniques are based on an *in vitro* bioassay technique. For more detailed information on the various methods, the reader is referred to "Sampling Microorganisms and Aeroallergens" (ACGIH, 1989b).

There are three general categories of sampling methods: Gravity samplers, inertial impactors, and filtration devices (Burge and Solomon, 1987). Gravitational collectors consist of a collection plate which contains a growth medium. Collection occurs via both gravity and inertial impaction, as the air over the plate is rarely still. This method is not a recommended technique because it is subject to both qualitative and quantitative errors (Burge and Solomon, 1987). For example large spores, which may be innocuous, are over estimated and smaller, potentially more significant organisms are under estimated. And further, as there is no air volume information, it is impossible to calculate atmospheric concentrations or estimate actual exposures (Burge and Solomon, 1987). Inertial impactors and filtration devices are the preferred types of collection devices. Table 4.12 summarizes the various devices which are described below.

Table 4.12   Samplers commonly used for collection of indoor bioaerosols

Sampler type	Sampling Principle of operation	Recommended rate (Lpm*)	Sample time
1. Slit impactor	Impaction on rotating or stationary plate	30–700 continuous	1–60 min or 7-day
2. Sieve impactor a. single-stage portable	Impaction on agar; "rodac" plate	90 or 185	0.5 or 0.3 min
b. single-stage impactor	Impaction on agar; 100 mm plates	28	1 min
c. two-stage impactor	Impaction on agar; 2–100 mm plates	28	1–5 min
3. Filter cassettes	Filtration	1–2	15–60 min or 8 h
4. High-volume filtration	Filtration	140-1400	5 min–24 h
5. High-volume electrostatic	Electrostatic collection into liquid	Up to 1000	Variable
6. All-glass impingers	Impingement into liquid	12.5	30 min
7. Centrifugal impactor	Impaction on agar; plastic strips	Approximately 40	0.5 min

* Lpm = liters per minute
Source: Guidelines for the Assessment of Bioaerosols in the Indoor Environment, ACGIH, 1989b. (With permission of the American Conference of Governmental Industrial Hygienists.)

Inertial impactors operate by either pulling air across a culture plate or by rotating the plate through the air at a constant speed. The types of devices included in this category are slit impactors, sieve impactors, and liquid impingers. Slit impactors operate by rotating a plate containing a culture medium. They are useful in collecting time discriminated samples, e.g., when sampling before and after the potential source is disturbed. They can collect particles down to 10 μm, which includes pollens, various fungi spores and allergen carriers. However, human pathogens are not collected efficiently (Burge and Solomon, 1987). An example of this type of device is the Rotorod sampler, made by Ted Brown Assoc., Palo Alto, Colorado.

Sieve impactors are also called cascade impactors. The operational principle is that air is drawn through a wide orifice, accelerating as it passes through the plates of smaller and smaller caliber. Culture plates are placed under each sieve plate and the microorganisms are impacted on them. Larger particles are collected in the upper plates and smaller organisms are collected in the lower plates. This method can be used to collect both bacteria and fungi. Although a six stage impactor is made, use of either the fifth or sixth stages alone will provide an integrated sample (ACGIH, 1989b). The two stage model, developed for particulate matter sampling, is intended to collect both non-respirable and respirable particles. This is of limited use for bioaerosols because some of the larger micro-organisms which impact in the nasal passages can elicit symptoms (ACGIH, 1989b). Andersen Samplers, Atlanta, Georgia is perhaps the most well-known manufacturer of cascade impactors. Another device, the Burkard spore trap, is made by Burkard Manufacturing Co., Rickmansworth, England.

Liquid impingers are another type of impaction device. They draw air through a liquid, trapping the particles. This technique is used for the collection of soluble materials (e.g., mycotoxins, antigens, endotoxins) and for bacteria and fungi which require gentle handling (Burge and Solomon, 1987). Impingers are also useful in areas where very high concentrations are expected (e.g., agricultural environments) because the collecting liquid can be diluted for culture (ACGIH, 1989b).

Another type of device in this category is the centrifugal impactor. It is easily portable and allows sampling with minimal disruption of the occupants. A potential drawback noted by Macher and First (1983) is that the instrument is calibrated by the manufacturer and this calibration may be inaccurate.

The final type of device, a high-volume electrostatic sampler, is very sophisticated. It is designed to collect particulate matter continuously from a large volume of air and deposit it into a small amount of liquid. The sampler uses an electrostatic field, a corona, to separate microorganisms from the air stream. The sampler is used for collecting bioaerosols which cause health effects at very low concentrations, e.g., viruses or protozoa (ACGIH, 1989b). This instrument is made by Sci-Med, Model M-3A. The manufacturer claims that it has a 70% collection efficiency for particles down to 0.1 μm.

Filtration can be conducted using low-volume or high-volume samplers. The low-volume samplers are quiet and can be worn as personal samplers. The filter collects particles down to and sometimes below its rated pore size. Following sampling, the filters are washed and the extract cultivated. Polycarbonate filters are used because they have a smooth surface. This method is useful for fungus spores, because they are resistant to drying. ACGIH (1989b) recommends using this technique in environments with very high concentrations.

The same technique can be used with high-volume samplers. However, these samplers are not recommended for the indoor environment because their sampling rate is high relative to the total volume of the space being sampled.

Following the collection of an air sample of bioaerosols, the media must be analyzed. Burge and Solomon (1987) describe five major analytical methods: culture, direct microscopy, bioassay, biochemical assay, and immunological assay.

For cultural assays, the organisms are collected on media which will support a good general growth of microorganisms. Specific media may be required for the growth and identification of certain types of microorganisms. The two general types of media used are "malt extract" agar for saprophytic organisms (derive nourishment from non-living matter), e.g., fungi (it will not support bacteria) and "Casein soy peptone" agar for bacteria and thermophilic actinomycetes (Burge et al., 1987). Incubation times and temperature differ according to the types of organisms expected to be found. For example, fungi on the malt extract agar should be incubated at 22 to 25 °C under fluorescent or ultraviolet light for three to seven days. The Casein soy peptone agar for bacteria should be incubated at room temperature for environmental organisms, 35 °C for human source organisms, and 55 to 56 °C for thermophilic actinomycetes. Colonies are then counted with a microscope, according to manufacturer's specification. Results are reported as colony forming units (CFU)/m^3. These researchers note that cultural assays will underestimate the total CFU for several reasons including non-conductive growing conditions (e.g., crowding, inadequate nutrition, or growth inhibitors produced by some organisms). In addition, the limited duration of viability of microorganisms means that at any one time, the fraction of microbial type that is viable can vary substantially.

Direct microscopy is used when the morphology of the particles can be used for identification. The method can be augmented with specific stains for certain types of bioaerosols (Burge and Solomon, 1987).

Bioassays are used most frequently for dust extracts, pollens, or spores to determine an individual's hypersensitivities. This is done by introducing microgram quantities of the relevant antigen into the skin and observing if an immune response is produced.

Biochemical assays are used for mycotoxins (which are produced by molds) that have a known molecular structure and can be characterized with a molecular sieve and paper chromatography (Burge and Solomon, 1987).

Immunologic assays are the newest of the analytical techniques for microbial aerosols. They are used to identify specific antigens from air samples or from bulk source samples, such as house dust (for cockroach antigen), fungus-related antigens, or contaminated materials. The method, described in Burge and Solomon (1987), involves adsorbing the antigen to a solid phase and adding human serum with antigen-specific antibodies. Tagged antibodies, e.g., radio-labelled with ^{125}I, from another species are added, and the binding between the solid phase antigen and the human serum antibody is related to the concentration of the antigen.

Analysis of microbiological samples which are collected from bulk sources may use similar techniques to those described above; however, the test for the house dust mite antigen is different. Sampling for house dust is generally performed using a vacuum. The dust can be analyzed using a variety of methods for different organisms: flotation for identification of anthropods, dilution and culture plates for bacteria and fungi, and chemical analysis for the excretion products of dust mites. This last technique, as described by Kniest (1987), measures guanine, the end product of dust mite excretion, colorimetrically using an azo dye.

4.14.3 Interpreting Indoor Concentrations

Comparing concentrations of bioaerosols across studies is of limited value due to the heterogeneous nature of microorganisms and the variety of sampling and analytical techniques. If an investigator considers it necessary to conduct air sampling for bioaerosols, it may be of use to compare levels of the specific type of organism analyzed to values reported in the literature. This must be done with caution because of the great variability in levels of bioaerosols.

The recommended approach for obtaining interpretable air sampling data is threefold: (1) sample the hypothetical source, if possible, (2) conduct air sampling in three different locations: the complaint area, a non-complaint area, and outdoors, and (3) if the complaints represent allergic reactions, conduct skin tests for the species or compound isolated from the air sampling to determine if the individuals have hypersensitive reactions to this compound or species.

Identifying the hypothetical source can be straightforward or require experienced detective work. Visible molds in carpet or microbial growth in HVAC humidifiers are obvious sources. If visible sources are not evident, the investigator may have to determine if the complaints of the individuals affected represent hypersensitivity reactions, i.e., the reactions represent an immune system response. Relating those systems to a probable source can be challenging.

Sampling in both the complaint and a control environment is critical. ACGIH (1989b) cites a general guideline that a situation can be considered unusual if the bioaerosol concentrations in the complaint area are an order of magnitude higher than in the control area or if the types of organisms are different. However, the presence of an organism or antigen or even elevated concentrations does not prove a causal relationship.

A bioassay (skin test) of the suspected organism or antigen on the hypersensitive individuals is required before an investigator can conclude with reasonable certainty that the bioaerosol exposure is contributing to the symptoms (ACGIH, 1989b).

In conclusion, sampling for bioaerosols requires a very specialized expertise. The investigator must have experience in selecting the appropriate sampling methodology and collection media, collecting a representative sample, using a qualified laboratory to perform the analyses, and interpreting the results. A great deal of care must be used in each of these areas in order to obtain meaningful results.

4.15 References

ACGIH (1989a). *Advances in Air Sampling*, American Conference of Governmental Industrial Hygienists, Cincinnati, OH, Lewis Publishers, Chelsea, MI.

ACGIH (1989b). *Guidelines for the Assessment of Bioaerosols in the Indoor Environment*, American Conference of Governmental Industrial Hygienists, Cincinnati, OH.

AIHA (1989). "Odor Thresholds for Chemicals with Established Occupational Health Standards," American Industrial Hygiene Association, Akron, OH.

ASHRAE (1985). *Fundamentals Handbook*. American Society of Heating, Refrigerating and Air Conditioning Engineers, Inc., Atlanta, GA.

ASHRAE (1989). "Ventilation for Acceptable Indoor Air Quality," Standard ANSI/ASHRAE 62–1989, American Society for Heating, Refrigerating and Air Conditioning Engineers, Atlanta, GA.

ASTM (1989). "Standard Method for Formaldehyde in Indoor Air (Passive Sampler Methodology)," Method No. D-5014, American Society for Testing and Materials, Philadelphia, PA.

Abu-Jarad, F., and Fremlin, J.H. (1982). "The Activity of Radon Daughters in High-Rise Buildings and the Influence of Soil Emanation," *Environ. International* **8**:37–43.

Akland, G., Johnson, T., and Hartwell, T. (1984). "Results of the Carbon Monoxide Study in Washington, D.C., and Denver, Colorado, in the Winter of 1982–83," EPA Report No. EPA-600/D-87-178.

Alonza, J., Cohen, B.L., Rudolph, H., Jow, H.N., and Frohlicher, J.O. (1979). "Indoor–Outdoor Relationships for Airborne Particulate Matter of Outdoor Origin," *Atmos. Environ.* **13**:55.

Amendola, A., and Hanes, N. (1984). "Characterization of Indoor Carbon Monoxide Levels Produced by the Automobile," in *Indoor Air* **4**:97–102, *Proceedings of the 3rd International Conference on Indoor Air Quality and Climate*, Swedish Council for Building Research, Stockholm.

Andersen, I. (1972). "Relationships Between Outdoor and Indoor Air Pollution," *Atmos. Environ.* **6**:275–278.

Badre, R., Guilleron, R., Abran, N., Bourdin, M., and Dumas, G. (1978). "Atmospheric Pollution by Smoking," *Annal. Pharm. Fr.* **36**:443–452.

Berglund, B., Berglund, U., Johansson, I., and Lindvall, T. (1984). "Mobile Laboratory for Sensory Air Quality Studies in Non-Industrial Environments," in *Indoor Air* **3**:467–472, *Proceedings of the 3rd International Conference on Indoor Air Quality and Climate*, Swedish Council for Building Research, Stockholm.

Biersteker, K., de Graaf, H., and Nass, C. (1965). "Indoor Air Pollution in Rotterdam Homes," *International J. Air and Water Poll.* **9**:343–350.

Billick, I.H., and Nagda, N.L. (1987). "Reaction Decay of Nitrogen Dioxide," *Proceedings of the 4th International Conference on Indoor Air Quality and Climate*, Institute for Water, Soil, and Air Hygiene, Berlin.

Broder, I., Corey, P., Cole, P., Mintz, S., Lipa, M., and Nethercott, J. (1986). "Health Status of Residents in Homes Insulated with Urea-Formaldehyde Foam," in *Indoor Air Quality in Cold Climates*, D. Walkinshaw (ed.), Air Pollution Control Association, Pittsburgh, PA.

Brunekreef, B., Boleij, J., Hoek, F., Lebret, E., and Noy, D. (1986). "Variation of Indoor Nitrogen Dioxide Concentrations Over a One-Year Period," *Environ. International* **12**:279–282.

Brunekreef, B., Smit, H., Biersteker, K., Boleij, J., and Lebret, E. (1982). "Indoor Carbon Monoxide Pollution in the Netherlands," *Environ. International* **8**:193–196.

Bruno, R.C. (1983). "Sources of Indoor Radon in Houses: A Review," *J. Air Poll. Control Assoc.* **33**:105–109.

Burge, H.A., and Solomon, W.R. (1987). "Sampling and Analysis of Biological Aerosols," *Atmospheric Environment* **21**:451–456.

Burge, H.A., Chatigny, M., Feeley, J., Kreiss, K., Morey, P., Otten, J. and Peterson, K. (1987). "Guidelines for Assessment and Sampling of Saprophytic Bioaerosols in the Indoor Environment," *Applied Industrial Hygiene* **5**(2):R10–R16.

Cain, W.S. (1976). "Olfaction and the Common Chemical Sense: Some Psychophysical Contrasts," in *Sensory Processes*, Vol. 1, p. 57.

Chaney, L.W. (1978). "Carbon Monoxide Automobile Emissions Measured from the Interior of a Traveling Automobile," *Science* **199**:1203–1204.

Clausing, P., Mak, J.K., Spengler, J.D., and Letz, R. (1984). "Personal NO2 Exposure of High School Students," in *Indoor Air* **4**:135–139, *Proceedings of the 3rd International Conference on Indoor Air Quality and Climate*, Swedish Council for Building Research, Stockholm.

Colwill, D., and Hickman, A. (1980). "Exposure of Drivers to Carbon Monoxide," *J. Air Poll. Control Assoc.* **30**:1316–1319.

Consensus Workshop (1984). "Report on the Consensus Workshop on Formaldehyde," *Env. Health Perspectives* **58**:323–381.

Contant, C.F., Gehan, B.M., Stock, T.H., and Holquin, A.H. (1986). "Estimation of Individual Ozone Exposures Using Microenvironmental Measurements," EPA Report No. 600/D-86/032.

Coutant, R.W., and Scott, D.R. (1982). "Applicability of Passive Dosimeters for Ambient Air Monitoring of Toxic Organic Compounds," *Env. Sci. Technol.* **16**:410–413.

Coviaux, F., Mouilleseaux, A., Festy, B., Thibaut, G., Geronimi, J., and Viellard, H. (1984). "Air Quality and Biological Controls of Workers Exposed in Working Premises Contiguous to an Urban Road-Tunnel," in *Indoor Air* **4**:129–133, *Proceedings of the 3rd International Conference on Indoor Air Quality and Climate*, Swedish Council for Building Research, Stockholm.

Dally, K.A., Hanrahan, L., and Woodbury, M. (1981). "Formaldehyde Exposure in Nonoccupational Environments," *Arch. of Environ. Health* **36**:277–284.

De Bortoli, M., Knoppel, H., Pecchig, E., Peil, A., Rogora, H., Schauenburg, H., Schlitt, H., and Vissers, H. (1985). "Measurements of Indoor Air Quality and Comparison with Ambient Air: A Study on 15 Homes in Northern Italy," Commission of the European Communities, NTIS PB86–136280.

De Bortoli, M., Knoppel, H., Pecchig, E., Peil, A., Rogora, H., Schauenburg, H., Schlitt, H., and Vissers, H. (1986). "Concentrations of Selected Organic Pollutants in Indoor and Outdoor Air in Northern Italy," *Environ. International* **12**:343–350.

Dement, J.M., Smith, N., Hickey, J., and Williams, T. (1984). "An Evaluation of Formaldehyde Sources Exposures and Possible Remidial Actions in Two Office Environments," in *Indoor Air* **3**:99–104. *Proceedings of the 3rd International Conference* *on Indoor Air Quality and Climate*, Swedish Council for Building Research, Stockholm.

Diemel, J.A., Brunekreef, B., Boleij, J.S., Biersteker, K., and Veenstia, S.J. (1981). "The Arnhem Lead Study II Indoor Pollution and Indoor/Outdoor Relationships," *Environ. Res.* **25**:449–456.

Dockery, D.W., and Spengler, J.D. (1977). "Personal Exposure to Respirable Particulates and Sulfates versus Ambient Measurements," Presented at the 70th Annual Meeting of the Air Pollution Control Association, Houston, Texas.

Dockery, D.W., Spengler, J.D., Reed, M.P., and Ware, J. (1981). "Relationships Among Personal, Indoor and Outdoor NO_2 Measurements," *Environ. International* **5**:101–107.

Dravnieks, A. (1975). "Evaluation of Human Body Odors, Methods and Interpretation," *J. of the Society of Cosmetic Chemists* **26**:551.

Duffee, R.A., Jann, P.R., Flesh, R.D., and Cain, W.S. (1980). "Odor/Ventilation Relationships in Public Buildings," *Proceedings of the 73rd Annual Meeting of the Air Pollution Control Association*, Montreal, Quebec, June 1980, Air Pollution Control Association, Pittsburgh, PA.

Eckmann, A.D., Dally, K.A., Hanrahan, L.P., and Anderson, H.A. (1982). "Comparison of the Chromotropic Acid and Modified Pararosaniline Methods for the Determination of Formaldehyde in Air," *Environ. International* **1–6**:159–168.

Fanger, P.O. (1987). "A Solution to the Sick Building Mystery," *Proceedings of the 4th International Conference on Indoor Air Quality and Climate*, Institute for Water, Soil, and Air Hygiene, Berlin.

Fecker, I., Hangartner, M., and Wanner, H. (1987). "Measurement of Carbon Dioxide of Indoor Air to Control the Fresh Air Supply," in *Indoor Air '87* **2**:635–639, *Proceedings of the 4th International Conference on Indoor Air Quality and Climate*, Institute for Water, Soil, and Air Hygiene, Berlin.

Fine, D.H. (1975). "Trace Analysis of Volatile N-Nitroso Compounds by Combined Gas Chromatography and Thermal Energy Analysis," *J. Chrom.* **109**:271–279.

Flachsbart, P.G. (1985). "The Effectiveness of Priority Lanes in Reducing Commuter Travel Time and Exposure to CO on a Honolulu Arterial," Presented at the 78th Annual Meeting of the Air Pollution Control Association, Detroit, Michigan.

Flachsbart, P.G., and Ott, W. (1984). "Field Surveys of Carbon Monoxide in Commercial Settings Using Personal Exposure Monitors," EPA Tech. Report No. EPA-600/4–84–019.

Flachsbart, P.G., and Ott, W. (1986). "A Rapid Method for Surveying CO Concentrations in High-Rise Buildings," *Environ. International* **12**:255–264.

Fletcher, R.A. (1984). "A Review of Personal/Portable Monitors and Samplers for Airborne Particles," *J. Air. Poll. Control Assoc.* **34**:1014–1016.

Garry, V.F., Oatman, L., Pleus, R., and Gray, D. (1980). "Formaldehyde in the Home: Some Environmental Disease Perspectives," *Minn. Med.* **63**:107–111.

Gebefugi, I., and Korte, F. (1984). "Indoor Contamination of Household Articles Through Pentachlorophenol and Lindane," in *Indoor Air* **4**:37–322, *Proceedings of the 3rd International Conference on Indoor Air Quality and Climate*, Swedish Council for Building Research, Stockholm.

General Electric Company (1972). "Indoor–Outdoor Carbon Monoxide Pollution Study," Final Report to U.S. EPA, EPA-RA-73–020, Contract CPA 70–77, Office of Research and Monitoring, U.S. Environmental Protection Agency, Washington, D.C.

George, A. (1986). "Instruments and Methods for Measuring Indoor Radon and Radon Progeny Concentrations," in *Indoor Radon, Proceedings of an APCA International Specialty Conference*, Air Pollution Control Association, Pittsburgh, PA.

Gesell, T.F. (1983). "Background Atmospheric ^{222}Rn Concentrations Outdoors and Indoors: A Review," *Health Physics* **45**:289–302.

Goldstein, I.F., Hartel, D., and Andrews, L.R. (1985). "Monitoring Personal Exposure to

Nitrogen Dioxide," Presented at the 79th Annual Meeting of the Air Pollution Control Association, Detroit, MI.

Goldstein, I.F., Hartel, D., Andrews, L.R., and Weinstein, A.L. (1986). "Indoor Air Pollution Exposures of Low-Income Inner City Residents," *Environ. International* 12:211–219.

Goldstein, B.D., Mellia, R.J., Chinn, S., Florey, C., Clark, D., and John, H. (1979). "The Relationship Between Respiratory Illness in Primary School Children and the Use of Gas Cooking II: Factors Affecting Nitrogen Dioxide Levels in the Home," *International J. Epi.* 8:339–345.

Good, B.W., Vilcins, G., Harvey, W.R., Clabo Jr., D.A., and Lewis, A.L. (1982). "Effect of Cigarette Smoking on Residential NO_2 Levels," *Environ. International* 8:167–176.

Hales, C.H., Rollinson, A.M., and Shain, F.H. (1974). "Experimental Verification of Linear Combination Model for Relating Indoor-Outdoor Pollutant Concentrations," *Environ. Sci. and Tech.* 8:452–453.

Halpern, M. (1978). "Indoor/Outdoor Air Pollution Exposure Continuity Relationships," *J. Air Poll. Control Assoc.* 28:689–691.

Hammond, S.K. and Leaderer, B.P. (1987). "A Diffusion Monitor to Measure Exposure to Passive Smoking," *Environ. Sci. and Tech.* 21:494–497.

Hammond, S.K., Leaderer, B.P., Roche, A.C., and Schenker, M. (1987). "Collection and Analysis of Nicotine as a Marker for Environmental Tobacco Smoke," *Atmos. Environ.* 21:457–462.

Hanrahan, L.P., Anderson, H., Dally, K., Eckmann, A., and Kanarek, M. (1985). "Formaldehyde Concentrations in Wisconsin Mobile Homes," *J. Air Poll. Control Assoc.* 35:1164–1167.

Hanrahan, L.P., Dally, K., Anderson, H., Kanarek, M., and Rankin, J. (1984). "Formaldehyde Vapor in Mobile Homes: A Cross Sectional Survey of Concentrations and Irritant Effects," *Am. J. Public Health* 74:1026–1027.

Harley, J.H. (1981). "Radioactive Emissions and Radon." *Bull. NY Acad. Med.* 57:883–896.

Hawthorne, A.R., Gammage R.G., and Dudney, C.S. (1986). "An Indoor Air Quality Study of 40 East Tennesse Homes," *Environ. International* 222–239.

Hawthorne, A.R., Gammage, R., Dudney, C., Mathews, T., and Eroman, D. (1984). "Formaldehyde Levels in Forty East-Tennessee Homes," in *Indoor Air* 3:17–22, *Proceedings of the 3rd International Conference on Indoor Air Quality and Climate*, Swedish Council for Building Research, Stockholm.

Hess, C.T., Weiffenback, C.V., and Norton, S.A. (1982). "Variations of Airborne and Waterborne Rn-222 in Houses in Maine," *Environ. International* 1–6:59–66.

Hildingson, O. (1982). "Radon Measurements in 12,000 Swedish Homes," *Environ. International* 8:67–70.

Hooper, J.E., and Cha, S. (1988). "Odor Perception and Its Measurement: An Approach to Solving Community Odor Problems," TRC Environmental Consultants, East Hartford, CT, U.S.A.

Hosein, R., Silverman, F., Corey, P, and Mintz, S. (1986). "The Relationship Between Pollutant Levels in Homes and Potential Sources," in *Indoor Air Quality in Cold Climates*, D. Walkinshaw (ed.), Air Pollution Control Assoc., Pittsburgh, PA.

Hümpel, P., and Keller, R. (1987). "Detection of Small Concentrations of Pentachlorophenol in Air," in *Indoor Air '87, Proceedings of the 4th International Conference on Indoor Air and Climate*, Berlin (West) 1:227–229.

Humphreys, M., Knight, C., and Pinnix, J. (1986). "Residential Wood Combustion Impacts on Indoor Carbon Monoxide and Suspended Particulates," *Proceedings from the 1986 EPA/APCA Symposium on the Measurement of Toxic Air Pollutants*.

Ingersoll, J.G. (1983). "A Survey of Radionuclide Contents and Radon Emanation Rates in Building Materials Used in the U.S.," Health *Physics* 45:363–368.

Jackson, M.D., and Lewis, R.G. (1981). "Insecticide Concentrations in Air after Application of Pest Control Strips," *Bull. Environ. Contam. Toxicol.* **27**:122–125.

James, A.C. (1987). "A Reconsideration of Cells at Risk and Other Key Factors in Radon Daughter Dosimetry," in *Radon and Its Decay Products*, P. Hopke (ed.), American Chemical Society, Washington, D.C.

Johansson, I. (1978). "Determination of Organic Compounds in Indoor Air with Potential Reference to Air Quality," *Atmos. Environ.* **12**:1371–1377.

Ju, C., and Spengler, J.D. (1981). "Room-to-Room Variations in Concentrations of Respirable Particles in Residences," *Environ. Sci. and Tech.* **15**:592–596.

Jurinski, N.B. (1984). "The Evaluation of Chlordane and Heptachlor Vapor Concentrations Within Buildings Treated for Insect Pest Control," in *Indoor Air* **4**:51–55, *Proceedings of the 3rd International Conference on Indoor Air Quality and Climate*, Swedish Council for Building Research, Stockholm.

Kelly, T.J., Barrier, R.H., and McClenny, W.A. (1989). "Real-Time Monitors for Formaldehyde in Ambient and Indoor Air," *Proc. of the 1989 EPA/AWMA International Symposium on Measurement of Toxic and Related Air Pollutants*, VIP-13, Air and Waste Management Assoc., Pittsburgh, PA.

Kim, Y.S., Spengler, J.D., and Yanagisawa, Y. (1986). "NO_2 Concentrations in Offices with Kerosene Space Heaters and Electric Stoves," in *Indoor Air Quality in Cold Climates*, D. Walkinshaw (Ed.), Air Pollution Control Association, Pittsburgh, PA.

Kniest, F. (1987). "Colorimetric Quantification of Inhalant Allergan Sources in House Dust," *Proceedings of the 4th International Conference on Indoor Air Quality and Climate*, Institute for Water, Soil, and Air Hygiene, Berlin.

Konopinski, V.J. (1984). "Residential Formaldehyde and Carbon Dioxide," in *Indoor Air* **3**:329–334, *Proceedings of the 3rd International Conference on Indoor Air Quality and Climate*, Swedish Council for Building Research, Stockholm.

Konopinski, V.J. (1989). "Residential Localized Carbon Dioxide Concentrations," in *Man and His Ecosystem* **1**:339–344, *Proceedings of the 8th World Clean Air Congress, The Hague, Netherlands*, Elsevier Science Publishers B.V., Amsterdam.

Koontz, M.D., and Nagda, N.L. (1986). "Exposures to Respirable Particulates: A Comparison of Instrumentation for Microenvironmental Monitoring," Report to the Electric Power Research Institute, EPRI Project RP2373–1, Palo Alto, CA.

Kothari, B.K. (1984). "Contribution of Soil Gas, Potable Water, and Building Material to Radon in U.S. Homes," in *Indoor Air* **2**:143–148, *Proceedings of the 3rd International Conference on Indoor Air Quality and Climate*, Swedish Council for Building Research, Stockholm.

Lamm, S.H. (1986). "Irritancy Levels and Formaldehyde Exposure in U.S. Mobile Homes," in *Indoor Air Quality in Cold Climates*, D. Walkinshaw (ed.) Air Pollution Control Association, Pittsburgh, PA.

Lauridsen, J., Muhaxheri, M., Clausen, G.H., and Fanger, P.O. (1987). "Ventilation and Background Odor in Offices," *Proceedings of the 4th International Conference on Indoor Air Quality and Climate*, Institute for Water, Soil, and Air Hygiene, Berlin.

Lawrence Berkeley Laboratory (1980). "Indoor Air Quality Measurements in Energy Efficient Buildings," LBL Report 8894, Lawrence Berkeley Laboratory, Berkeley, California.

Leaderer, B.P., Zagraniski, R.T., Berwick, M., Stolwijk, J.A.J., and Quing-Shan, M. (1984). "Residential Exposures to NO_2, SO_2, and HCHO Associated with Unvented Kerosene Heaters, Gas Appliances and Sidestream Tobacco Smoke," in *Indoor Air* **4**:151–156, *Proceedings of the 3rd International Conference on Indoor Air Quality and Climate*, Swedish Council for Building Research, Stockholm.

Lebret, E. (1985). "Air Pollution in Dutch Homes: An Exploratory Study in Environmental

Epidemiology," Ph.D. Thesis, Wageningen Agriculture University, the Netherlands, Report R-138, Report 1985–221.

Lebret, E., van de Wiel, H., Bos, H., Noy, D., and Boleij, J. (1986). "Volatile Organic Compounds in Dutch Home," *Environ. International* **12**:323–332.

Levin, J., Lindahl, R., and Anderson, K. (1989). "Monitoring of Parts-Per- Billion Levels of Formaldehyde Using a Diffusive Sampler," *J. Air Pollution Control Assoc.* **39**:44–47.

Lewis, R.G. (1989). "Development and Evaluation of Instrumentation for Measurement of Indoor Air Quality," *Man and His Ecosystem, Proceedings of the 8th World Clean Air Congress 1989*, The Hague, Elsevier, Amsterdam.

Lewis, R.G., and MacLeod, K.E. (1982). "Portable Sampler for Pesticides and Semi-Volatile Industrial Organic Chemicals in Air," *Anal. Chem.*, **54**:310–315.

Lewis, R.G., and Mulik, J.D. (1989). "Recent Developments in Passive Sampling Devices," in *Advances in Air Sampling*, American Conference of Governmental Industrial Hygienists, Cincinnati, OH, Lewis Publishers, Chelsea, MI.

Lewis, R.G., Bond, A.E., Fitz-Simons, T.R., Johnson, D.E., and Hsu, J.P. (1986). "Monitoring for Non-Occupational Exposure to Pesticides in Indoor and Personal Respiratory Air," Presented at the 79th Annual Meeting of the Air Pollution Control Association, Minneapolis, MN.

Lin, C.I., Anaclerio, R., Anthon, D., Fanning, L., and Hollowell, C. (1979). "Indoor/Outdoor Measurements of Formaldehyde and Total Aldehydes," presented at the 178th National Meeting of the American Chemical Society, Division of Environmental Chemistry, Washington, DC.

Lioy, P.J., Watson, J.G., and Spengler, J.D. (1980). "APCA Specialty Conference Workshop on Baseline Data for Inhaled Particulate Matter," *J. Air Poll. Control Assoc.* **30**:1126–1140.

Lipari, F., and Swain, S.J. (1985). "2,4-Dinitrophenylhydrazine-Coated Florisil Sampling of Formaldehyde in Air," *Env. Sci. Technol.* **19**:70–74.

Livingston, J.M., and Jones, C.R. (1981). "Living Area Contamination by Chlordane Used for Termite Treatment," *Bull. Environ. Contam. Toxicol.* **27**:406–411.

Lodge, J.P. Jr. (ed.) (1988). *Methods of Air Sampling and Analysis*, Lewis Publishers, Chelsea, Michigan.

Macher, J., and First, M. (1983). "Reuter Centrifugal Air Sampler: Measurement of Effective Airflow Rate and Collection Efficiency," *Appl. Environ. Microbiol.* **45**:1960–1964.

Marchant, R., Yoshida, K., and Figley, D., (1986). "Exposure to a High Concentration of Formaldehyde—A Case Survey," in *Indoor Air Quality in Cold Climates*, D. Walkinshaw (ed.), Air Pollution Control Association, Pittsburgh, PA.

Martell, E.A. (1984). "Aerosol Properties of Indoor Radon Decay Products," in *Indoor Air* **3**:161–166, *Proceedings of the 3rd International Conference on Indoor Air Quality and Climate*, Stockholm.

Melcher, R.G., Garner, W.L., Severs, L.W., and Vaccaro, J.R. (1978). "Collection of Chlorpyrifos and Other Pesticides in Air on Chemically Bonded Sorbents," *Anal. Chem.* **50**:25–255.

Meyer, B. (1983). *Indoor Air Quality*, Addison-Wesley Publishing Co., Inc., Reading, MA.

Miksch, R., Anthon, D., Fanning, L., Revzan, K., Glanville, J., and Hollowell, C. (1981). "The Modified Pararosaniline Method for the Determination of Formaldehyde in Air," *Anal. Chem.* **53**:2118–2123.

Miller, D.P. (1987). "Analysis of Nitrate in NO_2 Diffusion Tubes Using Ion Chromatography," U.S. Environmental Protection Agency Report No. 6000/4–87–013, U.S. Environmental Protection Agency, Cincinnati, OH.

Miller, F.J., Gardner, D.E., Graham, J.A., Lee, R.E., Wilson, W.E., and Bachman, J.D.

(1979). "Size Considerations for Establishing a Standard for Inhalable Particles," *J. Air Poll. Control Assoc.* **29**, 610.

Minamino, O. (1987). "Investigation Study between Air Contamination and Odor Sensation in the Office Building," *Proceedings of the 4th International Conference on Indoor Air Quality and Climate*, Institute for Water, Soil, and Air Hygiene, Berlin.

Mølhave, L., and Moller, J. (1979). "The Atmospheric Environment in Modern Danish Dwellings-Measurements in 39 Flats," in *Indoor Climate* 171–185, *Proceedings of the 1st International Indoor Climate Symposium*, Danish Building Research Institute, Copenhagen.

Moncrieff, R.W. (1967). *The Chemical Senses*, Leonard Hill, Div. International Textbook Co., Ltd., London, England.

Moschandreas, D.J., Stark, J., McFadden. J.E., and Morse, S.S. (1978). "Indoor Air Pollution in the Residential Environment," Geomet Report, Final Report to the U.S. EPA, Report No. 600/7-78/229, Vols I and II, U.S. Environmental Protection Agency, Washington, D.C.

Moschandreas, D.J., Winchester, J.W., Nelson, J.W., and Burton, R.M. (1979). "Fine Particle Residential Indoor Air Pollution," *Atmos. Environ.* **13**:1413–1418.

Moschandreas, D.J., Zabransky, J., and Rector, H.E. (1980). "The Effects of Woodburning on the Indoor Residential Air Quality," *Environ. International* **4**:463–468.

Moschandreas, D.J., and Rector, H.E. (1982). "Indoor Radon Concentrations," *Environ. International* **8**:77–82.

Moschandreas, D. and Zabransky, J. (1982). "Spatial Variation of Carbon Monoxide and Oxides of Nitrogen Concentrations Inside Residences," *Environ. International* **8**:177–183.

Moschandreas, D.J., Relwani, S.M., Taylor, K.C., and Mulik, J.D. (1989). "Performance of Nitrogen Dioxide Passive Sampling Devices Under Varying Climatic Conditions," *Proc. of the 1989 EPA/AWMA International Symposium on Measurement of Toxic and Related Air Pollutants*, VIP-13, Air and Waste Management Assoc., Pittsburgh, PA.

Mueller, F.X., Loeb, L., and Maper, W.H. (1973). "Decomposition Rates of Ozone in Living Areas," *Environ. Sci. and Technol.* **7**:347.

Nagda, N.L., Rector, H.E., and Koontz, M.D. (1987). *Guidelines for Monitoring Indoor Air Quality*, Hemisphere Publishing Corporation, New York, NY.

National Academy of Sciences (NAS) (1982). Committee on Toxicology, "An Assessment of Health Risks of Seven Pesticides Used for Termite Control," Washington, D.C.

National Academy of Sciences (1986a). *Environmental Tobacco Smoke: Measuring Exposures and Assessing Health Effects*, National Academy Press, Washington, D.C.

National Academy of Sciences (1986b). *The Airliner Cabin Environment*, National Academy Press, Washington, D.C.

National Institute of Occupational Safety and Health (1977). *NIOSH Manual of Analytical Methods*, 2nd edn. and Supplements, National Institute of Occupational Safety and Health, Cincinnati, OH.

National Institute of Occupational Safety and Health (1984). *NIOSH Manual of Analytical Methods*, 3rd Ed. and Supplements, National Institute of Occupational Safety and Health, Cincinnati, OH.

Nazaroff, W.W., and Nero, A.V. (1984). "Transport of Radon from Soil Into Residences," in *Indoor Air* **3**:15–20, *Proceedings of the 3rd International Conference on Indoor Air Quality and Climate*, Swedish Council for Building Research, Stockholm.

Nazaroff, W.W., and Nero, A.V. (eds) (1988). *Radon and Its Decay Products in Indoor Air*, John Wiley & Sons, New York, NY.

Nero, A.V. (1983a). "Airborne Radionuclides and Radiation in Buildings: A Review," *Health Physics* **45**:303–322.

Nero, A.V. (1983b). "Indoor Radiation Exposure from ^{222}Rn and Its Daughters: A View of the Issue," *Health Physics* **45**:277–288.

Nero, A.V. (1985). "Indoor Concentrations of ^{222}Radon and its Daughters: Sources Range and Environmental Influences" in *Indoor Air and Human Health*, R.B. Gammage and W.V. Kage (eds), Lewis Publishers, Chelsea, MI.

Nero, A.V. Schweher, M.B., Nazaroff, W.W., and Revan, K.L. (1986). "Distribution of Airborne ^{222}Radon Concentrations in U.S. Homes," *Science* **234**:992–993.

Norsted, S.W., Kozinetz, C., and Annegers, J. (1985). "Formaldehyde Complaint Investigations in Mobile Homes by the Texas Dept. of Public Health," *Environ. Research* **37**:93–100.

Noy, D., Lebret, E., Willers, H., Winkes, A., Boleij, J., and Brunekreef, B. (1986). "Estimating Human Exposure to Nitrogen Dioxide: Results From a Personal Monitoring Study Among Housewives," *Environ. International* **12**:407–411.

Oehme, M., and Knoppel, H. (1987). "Analysis of Low Volatile ($>C_{15}$) and Particle Bound Indoor Pollutants: Assessment of a Sensitive Method and First Result," in *Indoor Air '87, Proceedings of the 4th International Conference on Indoor Air Quality and Climate*, Berlin (West). Volume I, pp. 210–214.

Offermann, F.J., Hollowell, C.D., Nazaroff, W.W., Roseme, G.D., and Rizzuto, J.R. (1982). "Low-Infiltration Housing in Rochester, New York: A Study of Air-Exchange Rates and Indoor Air Quality," *Environ. International* **8**:435–446.

Ott, W.R., Rodes, C.E., Drago, R.J., Williams, C., and Burman, F.J. (1986). "Automated Data-Logging Personal Exposure Monitors for Carbon Monoxide," *J. Air Pollution Control Assoc.* **36**(8):883–887.

Palmes, E.D., Gunnison, A.F., DiMattio, J., and Tomczyk, C. (1976). "Personal Sampler for Nitrogen Dioxide," *Am. Ind. Hyg. Assoc. J.* **37**, 570.

Petersen, G., and Sabersky, R. (1975). "Measurements of Pollutants Inside an Automobile," *J. Air Poll. Control Assoc.* **25**:1028–1032.

Prichard, H.M., Gesell, T.F., Hess, C.T., Weiffenbach, C.V., and Nyberg, P. (1982). "Associations Between Grab Sample and Integrated Radon Measurements in Dwellings in Maine and Texas," *Environ. International* **8**:83–87.

Quackenboss, J.J., Kanarek, M.S., Spengler, J.D. and Letz, R. (1982). "Personal Monitoring for Nitrogen Dioxide Exposure: Methodological Considerations for a Community Study," *Environ. International* **8**:249–258.

Repace, J.L., and Lowery, A.H. (1980). "Indoor Air Pollution, Tobacco Smoke, and Public Health," *Science* **208**:464–572.

Repace, J.L., and Lowery, A.H. (1982). "Tobacco Smoke, Ventilation, and Indoor Air Quality," *ASHRAE Trans.* **88**:894–914.

Ryan, P.B., Spengler, J.D., and Letz, R. (1983). "The Effects of Kerosene Heaters on Indoor Pollutant Concentrations: A Monitoring and Modeling Study," *Atmos. Env.* **7**:1339–1345.

Rylander, R. (1985). "Workshop Perspectives," *Environ. J. Resp. Dis.* **65**:143–146.

Sabersky, R H , Sinema, D.A., and Shara, F.H. (1973). "Concentrations, Decay Rates, and Removal of Ozone and Their Relation to Establishing Clean Indoor Air," *Environ. Sci. and Tech.* **4**:347–353.

Sachs, H.M., Hernandez, T.L., and Ring, J.W. (1982). "Regional Geology and Radon Variability in Buildings," *Environ. International* **8**:97–103.

Schiff, H.I., MacKay, G.I., Castledine, C., Harris, G.W., and Tran, Q. (1986). "A Sensitive Direct Measurement NO_2 Instrument," *Proc. 1986 EPA/APCA Symposium on Measurement of Toxic Air Pollutants*, Raleigh, NC, pp. 834–844.

Schutte, W.C., Frank, C., and Long, K. (1981). "Problems Associated with the Use of Urea-Formaldehyde Foam for Residential Insulation," Contract No. W-7405-eng-26, Oak Ridge National Laboratory.

Sega, K., Kalimic, N., and Sisovic, A. (1984). "Indoor–Outdoor Relationships for Respirable Particles, Total Suspended Particulate Matter and Smoke Concentrations in Modern Office Buildings," in *Indoor Air* **2**:189–193, *Proceedings of the 3rd International Conference of Indoor Air Quality and Climate*, Swedish Council for Building Research, Stockholm.

Seifert, B., and Abraham, H.J. (1982). "Indoor Air Concentrations of Benzene and Some Other Aromatic Hydrocarbons," *Ecotoxicology and Environ. Safety* **6**:190–192.

Seifert, B., and Schmahl, H.J. (1987). "Quantification of Sorption Effects for Selected Organic Substances Present in Indoor Air," in *Indoor Air '87, Proceedings of the 4th International Conference on Indoor Air and Climate*, Berlin (West), Volume 1, pp. 252–256.

Sexton, K., Petreas, M., Liv, K., and Kulasingam, G. (1985). "Formaldehyde Concentrations Measured in California Mobile Homes," Presented at the 78th Annual Meeting of the Air Poll. Control Assoc., Detroit, MI.

Sexton, K., Spengler, J.D., and Treitman, R.D. (1984). "Effects of Residential Wood Combustion on Indoor Air Quality: A Case Study in Waterbury, Vermont," *Atmos. Environ.* **7**:1371–1383.

Sextro, R.G., Moed, B.A., Nazaroff, W.W., Revzan, K.L., and Nero, A.V. (1987). "Investigations of Soil as a Source of Indoor Radon," in *Radon and Its Decay Products*, P. Hopke (ed.), American Chemical Society, Washington, D.C.

Sheldon, L.S., Sparacine, C.M., and Pellizzari, E.D. (1984). "Review of Analytical Methods for Volatile Organic Compounds in the Indoor Environment," in *Indoor Air and Human Health*, R.B. Gammage and S.V. Kaye (eds.), Lewis Publishers, Chelsea, MI.

Sheldon, L.S., Handy, R.W., Hartwell, T., Whitmore, R., Zelon, H., and Pellizzari, E.D. (1988), "Indoor Air Quality in Public Buildings, Volume I," EPA-600/S6-88/009a, U.S. Environmental Protection Agency, Washington, DC.

Shields, H.C., and Weschler, C.J. (1987). "Analysis of Ambient Concentrations of Organic Vapors with a Passive Sampler," *JAPCA* **37**:1039–1045.

Sisovic, A., and Fugas, M. (1985). "Indoor Concentrations of Carbon Monoxide in Selected Urban Microenvironments," *Environ. Monitoring and Asses.* **5**:199–204.

Spengler, J.D., Ferris, B.G., Dockery, D.W., and Speizer, F.E. (1979). "Sulfur Dioxide and Nitrogen Dioxide Levels Inside and Outside Homes and the Implications on Health Effects Research," *Environ. Sci. and Tech.* **13**:1276–1280.

Spengler, J.D., Dockery, D.W., Turner, W.A., Wolfson, J.M., and Ferris, B.G. (1981). "Longterm Measurements of Respirable Sulfates and Particles Inside and Outside Homes," *Atmos. Environ.* **15**:23–30.

Spengler, J.D., Duffy, C., Letz, R., Tibbits, T., and Ferris, B.G. (1983). "Nitrogen Dioxide Inside and Outside 137 Homes and Implications for Ambient Air Quality Standards and Health Effects Research," *Environ. Sci. and Tech.* **17**:164–168.

Spicer, C.W., Contant, R.W., Ward, G.F., Joseph, D.W., Gaynor, A.J., and Billick, I.H. (1987). "Rates and Mechanisms of NO_2 Removal from Indoor Air by Residential Materials," *Proc. the 4th International Conference on Indoor Air Quality and Climate*, Institute for Water, Soil, and Air Hygiene, Berlin.

Sterling, D.A., Stock, T., and Monteith, D. (1984). "Factors Influencing Formaldehyde Levels in Manufactured Housing," in *Indoor Air* **3**:139–148, *Proceedings of the 3rd International Conference of Indoor Air Quality and Climate*, Swedish Council for Building Research, Stockholm.

Sterling, D.A., Stock, T., and Monteith, D. (1986). "Factors Influencing Formaldehyde Levels in Manufactured Housing," in *Indoor Air Quality in Cold Climates*, D. Walkinshaw (Ed.), Air Pollution Control Association, Pittsburgh, PA.

Stock, T.H., Kotchman D.J., Contant, C.F., Buffler, P.A., Holquin, A., Gehan, B.M., and Noel, L.M. (1985). "The Estimation of Personal Exposures to Air Pollutants for a

Community-Based Study of Health Effects in Asthmatics—Design and Results of Air Monitoring," *J. Air Pollution Control Assoc.* **35**:1266–1273.

Stock, T.H. and Mendez, S. (1985). "A Survey of Typical Exposures to Formaldehyde in Houston Area Residents," *Am. Ind. Hyg. Assoc. J.* **46**:313–317.

Sullivan, J.L., Shirtliffe, C., and Svec, J. (1986). "Seasonal Variations in Formaldehyde Concentrations in Homes Insulated with Urea-Formaldehyde Foam," in *Indoor Air Quality in Cold Climates*, D. Walkinshaw (ed.) Air Pollution Control Association, Pittsburgh, PA.

Sutton, D.J., Nodoff, K.M., and Makino, K.K. (1978). "Ozone Concentrations in Residential Structures," *ASHRAE Journal* **18**:21–26.

Syrotynski, S. (1986). "Prevalence of Formaldehyde Concentrations in Residential Settings," in *Indoor Air Quality in Cold Climates*, D. Walkinshaw (ed.), Air Pollution Control Association, Pittsburgh, PA.

Syversen, T., Eide, I., and Malvik, B. (1985). "Chemical Air Quality in Energy-Efficient Houses," *SINTEF Report* STF21 A83101.

Tejada, S. (1986). *Int'l J. Environ. Anal. Chem.* **26**:167–185.

Thomas, T.C., and Seiber, J.N. (1974). "Chromsorb 102, an Efficient Medium for Trapping Pesticides from Air," *Bul. Env. Contam. & Tox.* **12**:17–25.

Thompson, C.V., Jenkins, R.A., and Higgins, C.E. (1989). "A Thermal Desorption Method for the Determination of Nicotine in Indoor Environments," *Environ. Sci. and Tech.* **23**:429–435.

Tosteson, T.D., Spengler, J.D., and Weker, R.A. (1982). "Aluminum Iron and Lead Content of Respirable Particulate Samples from a Personal Monitoring Study," *Environ. International* **8**:265–268.

Traynor, G., Apte, M., Dillworth, J., and Grimsrud, D. (1982). "Indoor Air Pollution from Portable Kerosene-Fired Space Heaters," Presented at Hazardous Heat: A Symposium on the Health and Safety of Kerosene Heaters, Hempstead, N.Y.

Turiel, I., Hollowell, C.D., Miksch, R.R., Rudy, J.V., and Young, R.A. (1983). "The Effects of Reduced Ventilation on Indoor Air Quality in an Office Building," *Atmos. Environ.* **17**:51–64.

TVA (1985). "Analysis of Indoor Air Quality Data from East Tennessee Field Studies," Oak Ridge National Laboratory, Oak Ridge, TN DE85–13213.

U.S. Code of Federal Regulations (1989). Part 50, U.S. Government Printing Office, Washington, D.C.

U.S. Environmental Protection Agency (1984). "Compendium of Methods for the Determination of Toxic Organic Compounds in Air," EPA-600/4–84–041.

U.S. Environmental Protection Agency (1986). *Air Quality Criteria for Lead Volumes I–IV*, Final EPA-600/8–83/028.

U.S. Environmental Protection Agency (1988a). *Research Needs in Human Exposure: A Comprehensive 5-Year Assessment (1989–1993)*, U.S. Environmental Protection Agency, Washington, D.C.

U.S. Environmental Protection Agency (1988b). *Compendium for Determination of Air Pollutants in Indoor Air*, Final Draft, U.S. Environmental Protection Agency, Research Triangle Park, NC.

U.S. Environmental Protection Agency (1989). "Determination of Respirable Particulate Matter in Indoor Air Using a Continuous Particulate Monitor," Compendium Method IP-10B, U.S. Environmental Protection Agency, Research Triangle Park, NC.

U.S. Surgeon General (1986). *Involuntary Smoking, A Report of the Surgeon General*, U.S. Department of Health and Human services.

Verschueren, K. (1983). *Handbook of Environmental Data on Organic Chemicals*. Van Nostrand Reinhold Co., New York.

Wade, W. A., Cot/, W.A., and Yocom, J.E. (1975). "A Study of Indoor Air Quality," *J. Air Pollution Control Assoc.* **25**:933–939.

Wallace, L.A. (1986). "Personal Exposures, Indoor and Outdoor Air Concentrations and Exhaled Breath Concentrations of Selected Volatile Organic Compounds Measured for 600 Residents of New Jersey, North Dakota, North Carolina and California," *Tox. and Environ. Chem.* **12**:215–236.

Wallace, L.A. (1987). *The Total Exposure Assessment Methodology (TEAM) Study*, EPA/600/6–87/002a, U.S. Environmental Protection Agency, Washington, D.C.

Wallace, L.A., (1987). Personal Communication with S.M. McCarthy.

Wallace, L.A., and Ott, W.R. (1982). "Personal Monitors: A State-of-the-Art Survey," *J. Air Poll. Control Assoc.* **32**:601–610.

Wallace, L.A., Pellizzari, E.D., and Gordon, S.M. (1985). "Organic Chemicals in Indoor Air: A Review of Human Exposure Studies and Indoor Air Quality Studies," in *Indoor Air and Human Health*, Lewis Publishers, Chelsea, MI.

Wanner, H.U., and Kuhn, M. (1984). "Indoor Air Pollution by Building Materials," in *Indoor Air* **3**:29–34, *Proceedings of the 3rd International Conference on Indoor Air Quality and Climate*, Stockholm.

Weber, A., and Fischer, T. (1980). "Passive Smoking at Work," *Int. Arch. Occup. Environ. Health* **47**:209–221.

Wendel, G.J., Stedman, D.H., Cantrell, C.A., and Damrauer, L. (1983). "Luminol-based Nitrogen Dioxide Detector," *Anal. Chem.* **55**:937–40.

Weschler, C.J., Shields, H.C., and Naik, D.V. (1989). "Indoor Ozone Exposures," *J. Air Poll. Control Assoc.* **39**:1562–1568.

Winberry, W.T., Jr., Murphy, N.T., and Corouna, B. (1988). "Compendium of Methods for Determination of Air Pollutants in Indoor Air," Contract No. 68–02–4467, U.S. Environmental Protection Agency, Research Triangle Park, NC.

Wilson, M.J.G. (1968). "Indoor Air Pollution," *Proceedings of the Royal Society of London*, **A300, 307**:215–221.

Wright, C.G., and Leidy, R.B. (1980). "Insecticide Residues in the Air of Buildings and Pest Control Vehicles," *Bull. of Environ. Contam. Toxicol.* **24**:582–589.

Wright, C.G., Leidy, R.B., and Dupree, H.E. (1981). "Insecticides in the Ambient Air of Rooms Following Their Application for Control of Pests," *Bull. Environ. Contam. Toxicol.* **26**:548–553.

Yarmac, R.F., McCarthy, S.M., and Yocom, J.E. (1987). "Final Report on an Unvented Gas Space Heater Study," prepared for the Consumer Product Safety Commission, Washington, D.C.

Yocom, J.E. (1982). "Indoor–Outdoor Air Quality Relationships: A Critical Review," *J. Air Poll. Control Assoc.* **32**:500–520.

Yocom, J.E., Clink, W.L., and Cot/, W.A. (1970). "A Study of Indoor-Outdoor Air Pollution Relationships," Vol I. Final Report prepared for the National Air Pollution Control Association, Washington, D.C.

Yocom, J.E., Clink, W.L., and Cot/, W.A. (1971). "Indoor/Outdoor Air Quality Relationships," *J. Air Poll. Control Assoc.* **21**:251–259.

Yocom, J.E., Baer, N.S., and Robinson, E. (1986). "Air Pollution Effects on Physical and Economic Systems" in *Air Pollution* **6**:145–246, Academic Press, New York.

Ziskind, R., Rogozen, M., Carlin, T., and Drago, R. (1981). "Carbon Monoxide Intrusion into Sustained-Use Vehicles," *Environ. International* **5**:109–23.

Ziskind, R., Fite, K., and Mage, D. (1982). "Pilot Field Study: Carbon Monoxide Exposure Monitoring in the General Population," *Environ. International* **8**:283–293, 1982.

CHAPTER 5

Measurement of Pollutant Emissions

Table 1.3 in Chapter 1 lists various pollutants and their indoor and/or outdoor sources. A full understanding of a given indoor air quality situation requires specific knowledge about the sources that contribute to the concentrations of pollutants measured indoors. The measurement of emissions from sources of outdoor pollutants will not be addressed. However, we must have some way of assessing the contribution of outdoor air pollutants to indoor concentrations of pollutants. The most straightforward method is to measure outdoor concentrations of the pollutant simultaneously with indoor measurements. When there are indoor sources of the pollutant, the emission rate of the pollutant should be measured. Such emission rate measurements provide extremely valuable information for:

- assigning the contribution to indoor concentrations of a given source relative to outdoor and other indoor sources,
- determining emission rates in relation to important variables such as environmental effects (e.g., effect of temperature on release of VOC from building materials) and human intervention (e.g., operation of a gas stove or use of a consumer product), and
- selecting methods of controlling emissions or reducing their effect on indoor air quality (e.g., determining the effectiveness of encapsulating materials that off-gas VOC or the effectiveness of range hoods in removing combustion products from unvented gas stoves).

As shown by equation (1.2) in Chapter 1, under steady state conditions indoor concentrations of an indoor-generated pollutant are a function of the indoor source strength, the outdoor concentration of the pollutant, the air exchange rate, and the indoor pollutant decay rate. This equation tells us that it should be possible to calculate indoor emission rates if we know the air exchange rate, indoor and outdoor concentrations of the pollutant in question, and the pollutant decay rate. However, such calculations provide only a rough indication of indoor emissions because of the variability in air exchange rates, variability in outdoor concentrations, and uncertainty about pollutant decay rates for reactive pollutants. For such pollutants (e.g., NO_2) decay rate depends on a variety of factors such as the physical and chemical characteristics of indoor surfaces (Billick and Nagda, 1987). Nevertheless,

Wallace (1989) has estimated indoor source strengths for a group of volatile organic compounds (VOC) based on indoor/outdoor concentrations, air exchange rates and the assumption that indoor decay rates for such compounds are zero. The most reliable approach is to set up *in situ* or laboratory systems to measure emissions from a source directly. This chapter is not intended to be an exhaustive review of all of the work to measure indoor pollutant emission rates, but rather is a review of typical examples of the most important configurations. Many examples of these methods are discussed in the results of a U.S. EPA workshop on "Characterization of Contaminant Emissions from Indoor Sources" (Leaderer, 1987).

5.1 Measurement Approaches

The following sections discuss emission measurement approaches in terms of three categories of pollutants: combustion pollutants, VOC, and radon. Before discussing specific approaches used to measure emissions of a given category of pollutant, it is advisable to point out briefly the various physical configurations used for such emission measurements.

5.1.1 Environmental Chambers

Except for an example described later, most environmental chambers are at fixed locations in laboratories. In other words, the indoor source must be brought to the laboratory and emissions measured within the environmental chamber. Such environmental chambers range in sizes from a fraction of a cubic meter to large chambers in which a number of humans can operate. Such chambers, regardless of size, are self-contained and provide a controlled environment in terms of temperature, humidity, relatively non-reactive surfaces and air quality (except for the contribution of the source being evaluated). Since the environment is to a degree synthetic, there are often questions about the representativeness of chamber test results in relation to real-life situations. The problem of extrapolating results from small chambers to full-size, real-life buildings is especially critical.

5.1.2 Temporary Hoods and Partial Enclosures

Temporary hoods and partial enclosures placed over an emitting source permit the collection of pollutant emissions *in situ*. Such configurations are usually limited to well defined sources such as unvented heaters and stove-top gas burners. The hood or enclosure is presumed to have no significant effect on the process by which pollutants are emitted from a source. In the case of combustion devices the hood or enclosure should not impede or enhance the normal flow of either primary or secondary combustion air nor should the sampling conditions in the off-take duct modify the chemical and physical conditions in the exhaust. The primary advantages of these configurations are their relative simplicity and low cost and their ability to be used *in situ*.

5.1.3 Headspace Analysis

Headspace analysis can take many forms. This approach is commonly used for determining emissions of pollutants from materials used indoors (e.g., adhesives, caulking compounds, coatings, carpeting and copier toners). The method is most applicable to measuring VOC emissions. The most common configuration is to place a known quantity of a material in a chamber such as a glass or stainless steel vial over which a stream of purified air or inert gas, such as nitrogen or argon, is passed. The temperature of the material and pressure over the material are controlled. Emissions are measured as weight emitted per weight of material or area of the material per unit over a given time period. Often the temperature of the material is increased above ambient, and the pressure is reduced to accelerate the loss of volatile or semi-volatile compounds.

Headspace analysis is an extremely useful method of screening materials for their potential to emit indoor pollutants, but one must be extremely careful in extrapolating such laboratory results to full-scale building situations.

5.2 Measuring Emissions from Combustion Sources

Combustion gases (i.e., NO, NO_2 and CO) have been measured by a number of workers. Coté *et al*. (1974) measured NO, NO_2 and CO emission rates from gas stoves and an unvented gas-fired space heater using a small (8.0 m^3) environmental chamber. The test chamber was also used to determine the effectiveness of over-the-stove hoods (both exhaust and recirculating) for removing pollutants. The chamber was constructed of gypsum wallboard coated on the inside with epoxy paint, and all joints around openings (door and observation windows) were sealed with a silicone sealant. Figure 5.1 is a schematic diagram of the chamber and the sampling system. The primary exhaust fan was capable of providing up to one air change per minute. In view of the rapid air change and the relatively unreactive epoxy coating on the walls of the chamber, NO_2 decay within the chamber was believed to be insignificant. As part of this same research, these workers measured NO_2 decay in a home containing reactive furnishings to be 0.83 h^{-1}.

Girman *et al*. (1982) of Lawrence Berkeley Laboratory (LBL) measured emission rates of CO, CO_2, RPM, HCHO, NO and NO_2 from combustion appliances and side stream cigarette smoke using a 27 m^3 environmental chamber and a Mobile Atmospheric Research Laboratory (MARL); the latter contained the continuous monitoring instruments. Emissions of respirable particulate matter and formaldehyde were measured by instruments located in the environmental chambers. Figure 5.2 is a diagram of the environmental chamber and the MARL

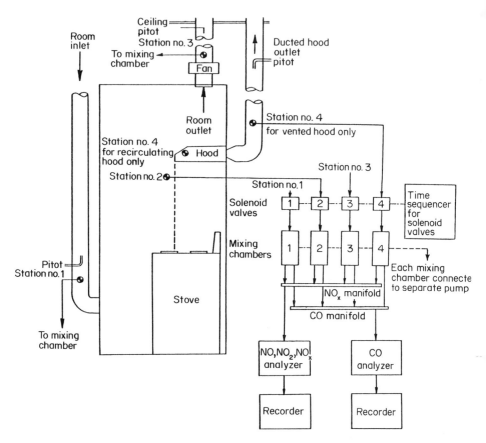

Figure 5.1 Schematic arrangement of environmental testing chamber and equipment arrangement. (Coté *et al*., 1974.) ◓, Pollutant concentration measurement points

used by this group. In measuring emissions from unvented gas space heaters, a "cold wall" composed of solar panels was used to help remove radiant heat.

Macriss *et al*. (1987) measured CO, NO_2, and NO emission gas space heaters in chambers using the LBL-recommended air exchange rate of 0.5 ACH plus two higher air exchange rates (4 and 10 ACH). They found that CO emission rates were higher at the low air exchange rate than at the higher two air exchange rates. A lesser effect was shown on NO emission rates, and NO_2 emissions were unaffected. This work indicates that in chamber tests to measure emissions from combustion sources, air exchange rates must be high enough to prevent air starvation in the combustion appliance.

Emissions from combustion appliances have been measured *in situ* by using inverted funnels and variations thereof to direct combustion products to the

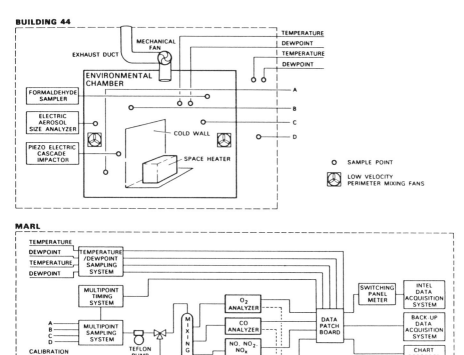

Figure 5.2   Schematic diagram of Lawrence Berkeley laboratory's environmental chamber and mobile atmospheric research laboratory (MARL). (Reproduced from Girman *et al*. (1982) with permission of Pergamon Press, Inc.)

sampling equipment. Such approaches have the advantage of permitting the testing of equipment as it would be operated under real-life conditions and not having to rely on laboratory-based environmental chambers.

Dave (1984, 1987) measured emissions of CO, NO, NO_2, SO_2, suspended particulate matter (SPM) and benzo-a-pyrene (BaP) from coal and kerosene-fired cook stoves of the types commonly used in India. He used a hood evacuated through a 100 mm diameter riser in which sampler inlets and flow measuring instruments were located. Emission rates were measured during periods when cooking utensils were over the flame and when they were absent.

Moschandreas *et al*. (1987) used a specially designed quartz dome placed over a range-top gas burner to facilitate the measurement of NO_2, NO_x and CO_2 emissions from a large number of gas stoves in homes located in the Chicago metropolitan area. Figure 5.3 is a schematic diagram of the experimental setup. The authors claimed that the quartz dome served the purpose of channelling, "concentrating," and mixing flue gases with excess air without disturbing the natural behavior of the open gas flame. A larger hood over the dome vented combustion products to the outside to prevent their buildup in the kitchen. Water vapor was removed from the sample stream before the sample was drawn into a Teflon™ bag. Samples were transported to a central laboratory in the Chicago area for analysis within 2 hours of the sampling time. NO/NO_x and CO_2 were measured using chemiluminescent and NDIR instruments, respectively.

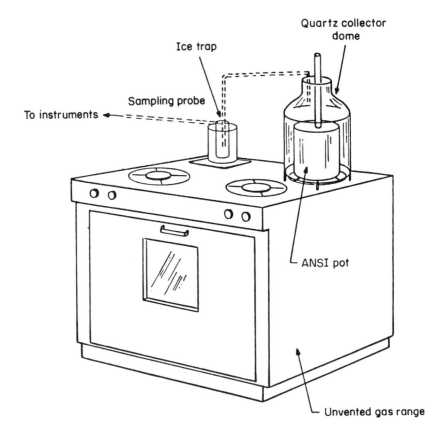

Figure 5.3   Experimental apparatus for measuring emission rates by the direct method. (Reproduced from Moschandreas *et al*. (1987) with permission of Pergamon Press, Inc.)

5.3 Measuring Emissions of Volatile Organic Compounds (VOCs)

VOCs (including formaldehyde) are emitted from countless indoor sources including consumer products, furnishings, and tobacco smoking. There is a large body of literature on emissions from tobacco smoking which has been summarized by the U.S. National Academy of Science (NAS, 1986). The methods used for measuring sidestream emissions from cigarette smoking have been reviewed in detail by Guerin et al. (1987). Therefore, there will be no attempt to cover the methods used for measuring emissions from tobacco smoking. This section of Chapter 5 will deal primarily with methods for measuring VOC emissions from such sources as construction materials and use of household products and non-combustion equipment.

Özkaynak et al. (1987) have reviewed the literature on sources and emissions of VOC from a large variety of construction materials and household products. The database they summarize was developed using a wide range of measurement techniques. Much of the data were from head-space analyses performed by the U.S. National Aeronautics and Space Administration (NASA) at reduced pressure and elevated temperature which would tend to produce extremely conservative emission rates. Nevertheless, the results of this review are useful in evaluating the relative emission potential for a wide range of materials.

Mølhave (1979) measured emissions of VOC from 32 common building materials used in Denmark. In later work (Mølhave, 1982), emissions from ten more materials were determined. The laboratory apparatus used is shown schematically in Figure 5.4. The environmental chamber is a 1 m^3 stainless steel box containing the material to be tested. The air through the chamber is continuously recirculated through the sampling tubes, particle filter, charcoal filters, and humidification system. VOCs in headspace air samples are collected on two series-mounted activated charcoal tubes. Analysis is by capillary column GC with detection by flame ionization or mass spectrometry. Sampling runs of different sample volumes are collected to identify the conditions of possible adsorbent breakthrough. The efficiency of the combined adsorption and desorption process is highly dependent on the type of compound collected. The method and the operating conditions restrict the range of molecular weights to between 50 and 200.

Girman et al. (1984) measured VOC emissions from adhesives. A total of 15 adhesives used for attaching floor tiles and ceiling and wall panels were sampled by vacuum extraction and cryogenic trapping of the emitted VOC. Collected samples were taken up in CS_2 and analyzed by GC/MS. In the next phase of the study, eight adhesives having the highest emission rates in the initial phase, after drying in accordance with the manufacturer's recommendations, were placed in a small temperature and humidity-controlled environmental chamber where VOC emission rates were measured using two-stage carbon adsorption tubes. The samples were desorbed with CS_2 and injected into a GC for flame ionization quantification.

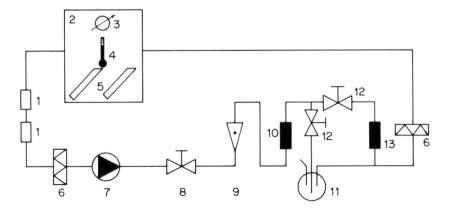

Figure 5.4   Recirculation system for air conditioning and air sampling of headspace over test pieces of building materials (adapted from Mølhave, 1979). 1, charcoal sampling tubes; 2, 1 m^2 hermetic sealed steel box; 3, hair hygrometer; 4, thermometer; 5, sample; 6, particle filter, 7, pump 0.69 l/min; 8, needle valve; 9, flowmeter; 10, charcoal filter; 11, glass bottle for air moistering; 12, on–off valve; 13, silica gel, air drier

The U.S. EPA has been studying VOC emissions from a wide range of materials used indoors (Tichenor, 1987). In this work modified incubators are being used as small environmental test chambers. Figure 5.5 is a schematic diagram of a typical chamber. Before a material is tested in the chamber, it is first screened through headspace analysis where effluents are characterized by GC/MS. In the environmental chamber temperature, relative humidity, and air exchange rates can be closely controlled. Air drawn from the chamber is collected on Tenax GC^{TM}/charcoal and analyzed by GC/FID. By taking a series of samples, the exponential decay of emissions can be determined. Figure 5.6 shows the decrease in time of the emission factor for a caulking compound in terms of three different VOCs. Even though the test chamber is constructed of polished stainless steel, the walls adsorb and desorb some of the emitted VOCs. Figure 5.7 illustrates the "sink effect" by showing the predicted mass of VOC adsorbed on the walls during the tests. These results demonstrate the importance of thorough cleaning of the chamber walls between tests.

Matthews *et al*. (1983) have developed a unique passive method for the measurement of formaldehyde emissions from building materials and indoor furnishings *in situ*. The sampler shown in Figure 5.8 consists of a 20.3 cm diameter mechanical sieve used as a sorbent support with a tight-fitting cover. The bottom surface of the sampler is covered with a soft gasket material. The sorbent used for formaldehyde measurement is 13X molecular sieve. The separation between the sorbent and the test surface (*a* in Figure 5.8) is 2.3 cm. Depending on anticipated formaldehyde emission rates, sampling periods vary between 1 and 3

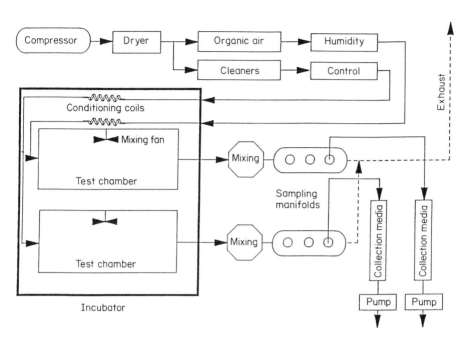

Figure 5.5   Schematic of example small chamber test facility (Tichenor, 1987)

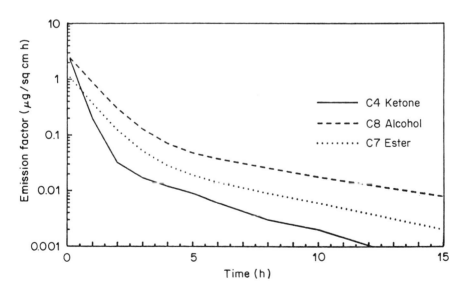

Figure 5.6   Caulk emissions versus time—three compounds (Tichenor, 1987). T, 23 °C; RH, 50%; ACH, 0.36

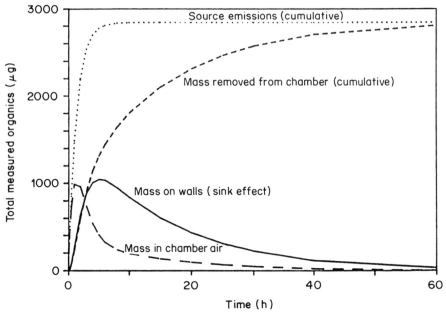

Figure 5.7   Chamber "sink effect"—caulking compound (Tichenor, 1987). T, 23 °C; RH, 50%; ACH, 0.36

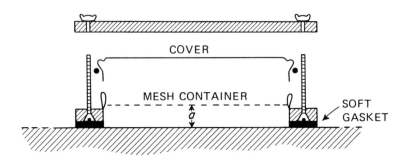

Figure 5.8   Formaldehyde surface emission monitor. (Reproduced from Matthews *et al.* (1983) with permission of the Air and Waste Management Association)

hours. Analysis of the adsorbed formaldehyde is by the pararosaniline colorimetric method. The authors claimed reasonably good agreement with results from chamber tests.

Kalmins and Gaudert (1986) used an apartment to measure formaldehyde emissions from particle-board and other indoor construction materials and furnishings. His approach was called "successive provisional elimination of sources" in which suspected sources of formaldehyde were removed or added

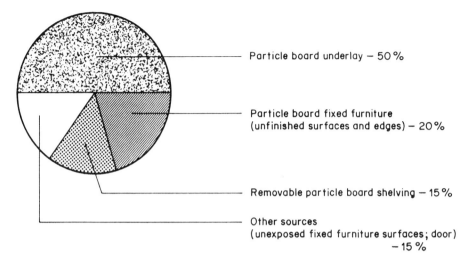

Particle board underlay – 50%

Particle board fixed furniture
(unfinished surfaces and edges) – 20%

Removable particle board shelving – 15%

Other sources
(unexposed fixed furniture surfaces; door)
– 15%

Figure 5.9   Estimated contributions of formaldehyde sources in the experimental apartment. (Reproduced from Kalmins and Gaudert (1986) with permission of the Air and Waste Management Association)

in accordance with a predetermined schedule. Figure 5.9 shows the relative contribution of various sources to indoor formaldehyde levels in this test apartment.

In a project to measure VOC emissions from various types of office equipment, one of the authors of this book designed and used a semi-portable environmental chamber that could be easily assembled and disassembled over various sizes and shapes of office machinery. Figure 5.10 is a schematic diagram of the chamber. The chamber was 2.4 m long, 1.5 m high, and 1.5 m wide. One of the side walls, ends and top were constructed of Tedlar^{TM} sheeting supported by lightweight aluminum angle. One side was constructed of sheet aluminum to provide the structural stability for mounting air handling and sampling equipment. Aluminum sheets were placed on the floor under the unit during testing to provide a non-reactive floor surface. Because of its relatively lightweight construction and ease of assembly and disassembly, the system could be taken to an office location for *in situ* measurement of equipment emissions.

Purified air was metered into the chamber at rates up to 3 to 5 ACH. Since the chamber was designed to leak because of intentional imperfect sealing and its positive pressure, and mixing fans provided relatively uniform concentrations of pollutants in the chamber, the emission rates of pollutants could be calculated from the in-chamber concentrations and the flow rate of purified air delivered to the chamber.

While emissions of criteria pollutants were measured, the primary emphasis was on the semi-continuous measurement of specific VOC using a TAGA 6000 Atmospheric Pressure Chemical Ionization (APCI)/MS/MS. The TAGA 6000

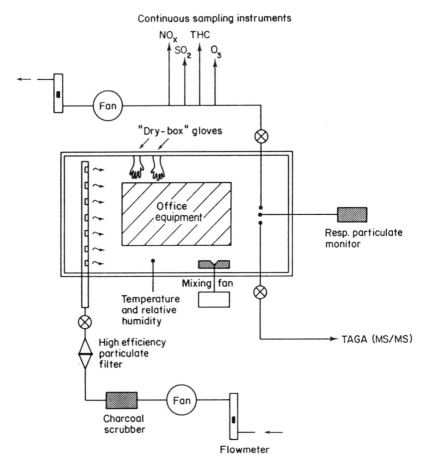

Figure 5.10   Schematic of test chamber for measuring emissions from office equipment

MS/MS consists of a triple quadropole mass spectrometer and an ultra high capacity cryogenic pump. The use of this device in indoor air quality investigations was described by Yocom *et al.* (1984).

5.4 Measuring Radon Emissions

As discussed in Chapter 4 (Section 4.7) radon as an indoor pollutant is not generated by discrete indoor sources such as NO_2 from combustion devices and VOC from building construction components or furnishings. The concentration of radon indoors is primarily a function of the rate at which it enters the indoor space from the soil surrounding or underneath a structure. Radon commonly enters a structure through various leakage paths such as cracks in a basement or slab floors through pipe penetrations. Other radon sources may be rock or concrete used in building foundations or well water from radium-containing formations. These other sources are not generally as important as the soil itself and will not be discussed further, therefore, this section will deal with unique methods for measuring the potential of soils to emit radon.

The radon source potential of soil depends on the release rate of radon from the soil and soil's permeability. In the method developed by Nazaroff and Sextro (1989), these two parameters can be quantified by measuring the flow rate and radon concentration of air drawn from the soil through a soil probe at various negative sampling pressures. The sampling equipment consists of probes made from galvanized pipe, diaphragm pressure gauges for measuring pressure differentials between 100 and 500 Pa, rotameters for measuring airflow rate between 1 and 3000 cm^3/min, a control valve, and a vacuum pump. In addition, a scintillation counter is required to measure radon concentration. The radon measurement system consists of an ~250 cm^3 flow through scintillation cell and a photomultiplier tube based on counting system. Figure 5.11 shows that the radon source potential measured in the soil by this method correlated well with radon concentrations in the basements of homes in New Jersey.

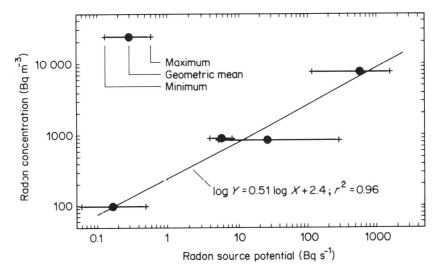

Figure 5.11   Indoor radon concentration measured in the basement versus radon source potential of soil measured at four New Jersey homesites. (Reproduced from Nazaroff and Sextro (1989) with permission of the American Chemical Society)

In a major program for the New York State Energy Research and Development Authority, Kunz (1989) used methods similar to those employed by Nazaroff and Sextro (1989) plus some additional techniques to determine the relationship between the radon-emitting potential of soils and indoor radon concentrations in several areas of New York State. Initially various areas of the state were surveyed by placing portable NaI μR meters directly on the soil to determine gamma emissions. Next, soil samples were collected to determine the content of a number of radionuclides. Grab samples to determine radon were then collected by techniques similar to those were used by Nazaroff and Sextro (1989). A schematic diagram of the equipment used is shown as Figure 5.12. In addition, time-integrated determinations of radon in soil gas were made using alpha track detectors in the device shown as Figure 5.13. Another important part of the program was to bury tracer gas sources between 1.5 and 2.0 feet from the foundations to test homes. The average fraction of emitted tracer entering the homes was 60%. This study found that by combining a measure of the radon concentration in soil gas with soil permeability, areas of the state can be characterized as to their potential to create indoor concentrations of radon. Figure 5.14 shows results from this study comparing soil gas radon with indoor radon concentrations for homes in different regions of New York State.

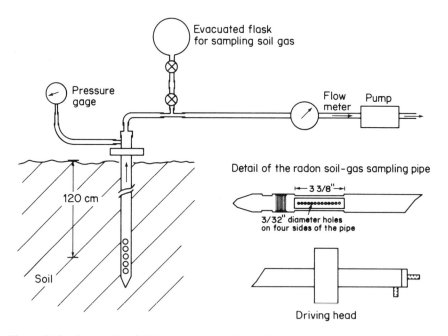

Figure 5.12   Schematic of field apparatus used to collect soil-gas samples and to measure the permeability of the soils for gas flow (Kunz, 1989)

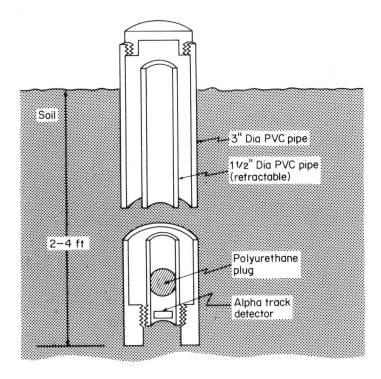

Figure 5.13   Schematic drawing of pvc pipe assembly used to make time-integrated soil-gas
Rn-222 measurements with alpha-track detectors (Kunz, 1989)

5.5 Conclusions

Measuring emissions of indoor air pollutants is an important component of any
indoor air quality measurement program. A variety of techniques is available
for making such measurements which range from those representing emissions
under nearly real-life conditions (e.g., large chambers or *in situ* hooding) to those
using synthetic conditions (e.g., headspace analysis carried out under reduced
pressure and/or elevated temperature). Regardless of the method used for measuring
emissions, in using such measurements to infer impact on indoor air quality one
must be extremely careful to assure that bias and inappropriate generalization does
not creep in. For example, measurements under laboratory conditions may not
indicate the extent to which pollutants decay or interact under real life conditions.

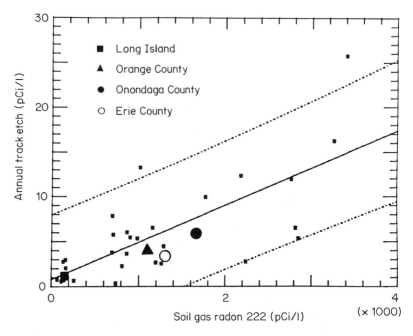

Figure 5.14   Soil gas Rn-222 plotted against indoor Rn-222 (Kunz, 1989)

References

Billick, I.H., and Nagda, N.L. (1987). "Reaction Decay of Nitrogen Dioxide," *Proceedings of the 4th International Conference on Indoor Air Quality and Climate,"* Institute for Water, Soil, and Air Hygiene, Berlin.

Coté, W.A., Wade, W.A., and Yocom, J.E. (1974). "A Study of Indoor Air Quality," EPA Contract 68-02-0745, U.S. Environmental Protection Agency, EPA 650/4-74-042, Washington, D.C.

Dave, J.M. (1984). "Studies on Emissions from Coal Burning Stoves (Sigries) as Used in Eastern India," *Proceedings of the 3rd International Conference on Indoor Air Quality and Climate,* Swedish Council for Building Research, Stockholm.

Dave, J.M. (1987). "Emissions from Conventional Kerosene Stoves Used in Indian Kitchens," *Proceedings of the 4th International Conference on Indoor Air Quality and Climate,* Institute for Water, Soil, and Air Hygiene, Berlin.

Girman, J.R., Apte, M.G., Traynor, G.W., Allen, J.R., and Hollowell, C.D. (1982). "Pollutant Emission Rates from Indoor Combustion Appliances and Sidestream Cigarette Smoke," *Environ. International* **8**:213–221.

Girman, J.R., Hodgson, A.T., Newton, A.S., and Winkes, A.W. (1984). "Volatile Organic Emissions from Adhesives with Indoor Applications," *Proceedings of the 3rd International Conference on Indoor Air Quality and Climate,* Swedish Council for Building Research, Stockholm.

Guerin, M.R., Higgins, C.E., and Jenkins, R.A. (1987). "Measuring Environmental Emissions from Tobacco Combustion: Sidestream Cigarette Smoke Literature Review," *Atmos. Environ.* **21**:291–297.

Kalmins, R., and Gaudert, P.C. (1986). "Formaldehyde Emissions from Typical Particle Board Applications and Assessment of Specific Abatement Measures," *Indoor Air Quality in Cold Climates*, D. Walkinshaw (ed.), Air Pollution Control Association, Pittsburgh, PA.

Kunz, C. (1989). "Influence of Surficial Soil and Bedrock on Indoor Radon in New York State Homes, Task II, Subtask II of an Investigation of Infiltration and Indoor Air Quality in New York State Homes," Report #89-14, New York State Energy Research and Development Authority, Albany, NY.

Leaderer, B.P. (ed.) (1987). "Characterization of Contaminant Emissions from Indoor Sources," *Atmos. Environ.* **21**:279–462.

Macriss, R.A., Elkins, R.H., and Billick, I.H. (1987). "Effect of Level of Air Infiltration on CO and NO Emissions from Unvented Space Heaters in Mass-Balance (Chamber) Measurements," *Proceedings of the 4th International Conference on Indoor Air Quality and Climate*, Institute for Water, Soil, and Air Hygiene, Berlin.

Matthews, T.G., Hawthorne, A.R., Daffron, C.R., and Reed, T.J. (1983). "Surface Emission Monitoring for Formaldehyde Source Strength Analysis," *Proceedings of Specialty Conference on Measurement and Monitoring on Non-Criteria (Toxic) Contaminants in Air*, Air Pollution Control Association, Pittsburgh, PA.

Mølhave, L. (1979). "Indoor Air Pollution Due to Building Materials," Proceedings of the First International Indoor Climate Symposium, Danish Building Research Institute, Copenhagen.

Mølhave, L. (1982). "Indoor Air Pollution Due to Organic Gases and Vapours of Solvents in Building Materials," *Environ. International* **8**:117–127.

Moschandreas, D.J., Relwani, S.M., Billick, I.H., and Macriss, R.A. (1987). "Emission Rates from Range-Top Burners—Assessment of Measurement Methods," *Atmos. Environ.* **21**:285–289.

National Academy of Sciences (1986). *Environmental Tobacco Smoke: Measuring Exposures and Assessing Health Effects*, National Academy Press, Washington, D.C.

Nazaroff, W.W., and Sextro, R.G. (1989). "Technique for Measuring the Indoor ^{222}Rn Source Potential of Soil," *Environ. Sci. Technol.* **23**:451–458.

Özkaynak, H., Ryan, P.B., Wallace, L.A., Nelson, W.C., and Behar, J.V. (1987). "Sources and Emission Rates of Organic Chemical Vapors in Homes and Buildings," *Proceedings of the 4th International Conference on Indoor Air Quality and Climate*, Institute for Water, Soil, and Air Hygiene, Berlin.

Tichenor, B.A. (1987). "Organic Emission Measurements via Small Chamber Testing," *Proceedings of the 4th International Conference on Indoor Air Quality and Climate*, Institute for Water, Soil, and Air Hygiene, Berlin.

Wallace, L.A. (1989). Personal Communication to J.E. Yocom.

Yocom, J.E., Hijazi, N.H., and Zoldak, J.J. (1984). "Use of Direct Analysis Mass Spectrometry to Solve Indoor Air Quality Problems," *Proceedings of the 3rd International Conference on Indoor Air Quality and Climate*, Swedish Council for Building Research, Stockholm.

CHAPTER 6

Indoor Air Quality Standards and Guidelines

In Chapter 1, outdoor ambient air quality standards were discussed. Such standards in the U.S.A. are legally enforceable, at least indirectly, by requiring control of the sources that emit the pollutants and then determining compliance with the air quality standards by air monitoring or by applying atmospheric diffusion models. Since the outdoor ambient atmosphere is in the public domain and there is a relatively clear cut relationship between the sources and the resulting ambient concentrations, such an approach is feasible and logical.

Such a straightforward approach in developing and applying indoor air quality standards is not possible. Indoor air quality is a function of both outdoor and indoor sources, building construction, air exchange rate and other factors, but one of the chief complications is that much of indoor exposure to pollutants is not in the public domain. Indoor air quality in the home, the most important microenvironment in defining total exposure to large segments of the population, may be dominated by domestic human activities such as smoking, cooking with unvented combustion devices, and using a wide range of consumer products. Control of indoor air quality in the domestic environment by the application of indoor air quality standards would be extremely difficult, if not totally infeasible.

In spite of these complications, the rapidly increasing concern about indoor air quality and its possible effects on health and well-being has fostered much activity internationally in the development of policies, guidelines and standards for indoor air quality. In the interest of brevity, these devices will be referred to as "standards." This area is developing so fast that this chapter can only be a snapshot in time (early 1990). At present there is a wide range of standards being used in a variety of ways to affect indoor air quality directly or indirectly. It is important for those conducting indoor air quality studies to know the current status of any indoor air quality policies, guidelines or standards in the regions of the world where they may be working.

The existence of indoor air quality policies, guidelines and standards implies that there are agencies or other entities responsible for maintaining indoor air quality at acceptable levels. These responsibilities are extremely fragmented and are shared, if taken at all, by a wide variety of organizations such as air pollution control, health, consumer product safety, and occupational hygiene agencies, plus agencies that utilize and enforce building and architectural codes.

Since indoor air quality standards do not yet exist in many parts of the world, those trying to solve indoor air quality problems invoke existing "standards," sometimes quite inappropriately. For example, occupational standards developed for 8-hour exposures of healthy workers in the industrial occupational environment are often utilized as the basis for evaluating exposure to pollutants in the office occupational environment. Another inappropriate use of occupational standards is to multiply the occupational standard (e.g., a threshold limit value (TLV) or permissible exposure limit (PEL)) by some fraction to make it, presumably, more applicable to the indoor non-industrial occupational setting. This practice is inappropriate because there are minimal or no experimental data to support these protection factors.

The most widely used "standard" in the U.S.A. is ASHRAE Standard 62-1989, "Ventilation for Acceptable Indoor Air Quality," published by the American Society for Heating, Refrigerating, and Air Conditioning Engineers (ASHRAE, 1989). While this "standard" does not have the force of law, it has been or is widely used as an enforceable standard in state and local building codes. This recently updated "standard" uses the U.S. ambient outdoor air quality standards for the quality of outdoor, fresh air make-up. It further specifies the quantity of outdoor make-up air per person for various building and occupancy categories. For example, the ventilation requirement for office buildings is 20 CFM/person. As stated in Chapter 4, this standard calls for a limit of 1000 ppm of CO_2 to satisfy comfort (odor) criteria. The standard includes a list of standards applicable in the U.S. for "common indoor air pollutants" and presents a list of indoor air pollutants of concern and their concentrations developed by the World Health Organization based on 1984 levels of knowledge (WHO, 1986). The WHO listing is presented here as Table 6.1.

The Air Infiltration and Ventilation Centre of the International Energy Agency has reviewed building airtightness standards in a number of countries. Table 6.2 shows the status of such standards as presented by Colthorpe (1990).

The National Institute of Building Sciences (NIBS, 1986) prepared a summary of indoor air quality standards, "Standards, Regulations, and Other Technical Criteria Related to Indoor Air Quality." The work on which this report is based is an extensive database from a wide range of international sources summarizing standards used or under development in many countries. It is hoped that this database will be kept up to date. These "standards" take many forms summarized as follows:

- Product oriented regulations.
  — Limitations on formaldehyde and other VOC emissions from building materials.
  — Limitations on use of asbestos containing materials.
  — Limitations on use of pesticides (e.g., chlordane).

Table 6.1 Consensus of concern about indoor air pollutants at 1984 levels of knowledge (WHO, 1986)

Pollutant*	Concentrations† reported	Concentrations† of limited or no concern	Concentrations† of concern	Remarks
Tobacco smoke (passive smoking)				
Respirable particulates	0.05–0.7	<0.1	>0.15	Japanese standard 0.15 mg/m^3
CO	1–1.5	<2	>5	Indicator for eye irritation (only from passive smoking)
Nitrosodimethylamine	$(1-50)\times10^{-6}$	—	—	Mutagens under investigation for carcinogenicity
NO_2	0.05–1	<0.19	>0.32	
CO	1–100	2% COHb	3% COHb	99.9%‡
		<11	>30	Continuous exposure
Radon and daughters	10–3000	≈0	70 Bq/m^3	Swedish standard for new houses
Formaldehyde	0.05–2	<0.06	>0.12	Long- and short-term
SO_2	0.02–1	<0.5	>1.35	SO_2 alone, short-term
CO_2	600–9000	<1800	>12000	Japanese standard 1800 mg/m^3
O_3	0.04–0.4	0.005	0.08	
Asbestos	<10 fibres/m^3	≈0	10^5 fibre/m	For long-term exposure
Mineral fibres	<10 fibres/m^3	—	—	Skin irritation

Organics

Methylene chloride	0.005–1	—	350	TLV§
			260	NIOSH‖ recommendations
Trichloroethene	0.0001–0.02	—	270	TLV
			135	NIOSH recommendations
Tetrachloroethene	0.002–0.05	—	335	TLV
1,4-Dichlorobenzene	0.005–0.1	—	450	TLV
Benzene	0.01–0.04	Carcinogen	Carcinogen	
Toluene	0.015–0.07	—	375	TLV
m,p-Xylene	0.01–0.05	—	435	TLV
n-Nonane	0.001–0.03	—	1050	ILO# (1980)
n-Decane	0.002–0.04	—	—	
Limonene	0.01–0.1	—	560	TLV turpentine

* All gases were considered on their own without other contaminants.

† Typical ranges of concentration given in mg/m^3, unless otherwise indicated, and for short-term exposures.

‡ According to Environmental Health Criteria No. 4, Geneva, World Health Organization, 1977.

§ TLV (threshold limit values) established by the American Conference of Governmental Industrial Hygienists (1983/1984). These values are for occupational exposures and should be considered as extreme upper limits for non-occupational populations for very short-term exposures.

‖ NIOSH: National Institute for Occupational Safety and Health, U.S.A.

ILO: International Labour Organization

— No meaningful numbers can be given because of insufficient knowledge.

AUTHORS' NOTE: Trichloroethene is now classed as a carcinogen by the U.S. EPA and the American Conference of Governmental Industrial Hygienists

Table 6.2 Requirements and recommendations for airtightness and ventilation rates in some countries*

	Scandinavia				Europe							America		Far East
	Den	Fin	Nor	Swe	Belg	Fra	Ita	Neth	Swi	UK	FRG	Can	USA	NZ
Airtightness:														
Components	W	W	R	N	W	—	W	W	W	W	W	W+D	W+D	W
Whole buildings	N	N	R	R	N	—	R	N[1]	R	N	N	N[1]	N[1]	N
Minimum ventilation rates:														
Dwellings	R	R	R	R	R[3]	R	R	R	N[2]	R	R	R	R	N[2]
Other (industrial/commercial)	R	R	R	R	—	—	R	R	R[5]	R	R	R	R	N[4]

Key: R   = Recommendation exists
N   = No recommendation exists
W   = Recommendation for windows only
W+D = Recommendation for doors and windows only

[1] Draft standard in preparation
[2] Recommendations exist for internal kitchens, bathrooms, toilets
[3] A voluntary standard that may soon be replaced
[4] Government legislation exists for bathrooms, toilets and laundries.
[5] Only for some types of rooms.

Country Abbreviations: Den Denmark; Fin Finland; Nor Norway; Swe Sweden; Belg Belgium; Fra France; Ita Italy; Neth Netherlands; Swi Switzerland; UK United Kingdom; FRG West Germany; Can Canada; USA United States of America; NZ New Zealand.

* Source: K. Colthorpe, "A Review of Building Airtightness and Ventilation Standards," Technical Note AIVC 30, Air Infiltration and Ventilation Centre, Coventry, Great Britain, September 1990. Reproduced with permission of the Air Infiltration and Ventilation Centre.

— Limitations on use of fiberglass or sound and thermal insulation inside of ducts.
— Limitations on leak rates (e.g., gas appliances).
— Specifications for performance of air-to-air heat exchangers.
• Smoking regulations.
— Total elimination of smoking.
— Designation of smoking areas.
• Building construction and maintenance standards.
— Restrictions on materials that may off-gas pollutants.
— Restrictions on sub-slab heating and ventilating ducts to avoid indoor contamination by termiticides.
— Methods to test leakage areas for windows.
— Standards for weatherization to minimize effects on indoor air quality.
— Standards for reducing exposure to lead from lead-based paint in building renovation and demolition.
— Standards for application of pesticides.
• Minimum ventilation requirements.
— Based on air changes per hour or air volume per occupant or unit areas for smoking and non-smoking areas.
— Based on indoor CO_2 concentrations.
— Requirements for windows and other methods of natural ventilation.
• Indoor air quality standards.
— Based on industrial hygiene standards (TLVs or PELs) or some fraction thereof.
— Based on outdoor air quality standards.
— Based on specific pollutants such as asbestos fibers, CO, VOC, formaldehyde, radon, and fluorocarbons.
— Limitations on CO_2 levels and degree of oxygen depletion (areas with combustion appliances).
• Standards for effectiveness of air cleaning devices.
— Specific methods for testing filter efficiency.
• Standards for conducting "sick building" investigations.

There are a wide range of organizations both public and private that have developed or are developing indoor air quality "standards." They include:

• National governmental organizations including those involved with the occupational and non-occupational environment, health and building construction.
• State, provincial, and city governments involved in these same areas.
• Government laboratories.
• Trade associations (e.g., gas appliance and construction materials).
• Technical societies (e.g., ASHRAE, ACGIH, etc.).

In addition to the database from NIBS, the World Health Organization (WHO) in Geneva compiles information on current indoor air quality legislation as part of the *International Digest of Health Legislation*.

As an example of the current status of "standards" development in one country, the U.S. EPA has recently designated indoor air quality as a high priority environmental problem (Mudarri, 1989). While the U.S. EPA is thus far reluctant to develop indoor air quality standards in the same mold as the outdoor ambient standards, this organization will pursue a program that will include both regulatory and non-regulatory methods for improving indoor air quality as described below:

(1) Non-regulatory methods to deal with the "sick building" type of problems:
— Information materials to the public concerning the causes, consequences, and approaches to solving the problem.
— Guidance, technical assistance and cooperative research with the building community.
— Training and development of enhanced capability in state and local governments, and private sector diagnostic firms.
— Cooperative ventures with the private sector to standardize diagnostic protocols.
(2) Regulatory and non-regulatory measures to deal with high risk sources and pollutants:
— Use of existing authority in U.S. EPA and other agencies to regulate toxic chemicals, pesticides, and consumer products.
— Development of information, fact sheets, and similar instruments on potential health impacts and problems of specific sources and pollutants, and recommended mitigation actions.
— Development of voluntary standards and guidelines for industry and the building community on pollutants and sources.
(3) Research programs in four areas:
— *Monitoring*: exposure monitoring, multipollutant field monitoring, and the development of personal and passive monitors.
— *Engineering*: source emission testing, model development, evaluation of air cleaners, and evaluation of control techniques.
— *Health effects*: methods for evaluating environmental tobacco smoke, effects on human subjects of volatile organic mixtures commonly found in buildings, and health effects of specific contaminants on sensitive populations (school children).
— *Risk Characterization*: evaluation of approaches to characterizing risk in support of policy decision-making.

In the U.S. each of the individual 50 states is considering indoor air quality in one way or another. California has a well developed indoor air quality group under the California Department of Health Services. As an example of another state program, the State of New Jersey under the Department of Labor has proposed indoor air quality standards for areas in which state employees work (Serraino, 1989). This proposed standard includes rigorous requirements for building owners and operators (where state employees work) to maintain records on such factors as the operation and maintenance of ventilation systems, air exchange rates, and filter efficiencies.

At this writing (early 1990), the U.S. Senate is considering a proposed bill entitled the "Indoor Air Quality Act of 1989" (U.S. Senate, 1989). This bill, if enacted, will provide for the following new programs including increased funding:

- Expanded research on indoor air quality.
- Assessment bulletins on technologies and management practices for control and measurement of indoor air quality.
- Indoor air contaminant health advisories.
- National air contaminant health advisories.
- Federal building response plan and demonstration program.
- Support of state indoor air quality programs.
- Establishment of an Office of Indoor Air Quality and a Council on Indoor Air Quality.
- Building assessment demonstration to support development of methods for assessing and controlling indoor air quality in buildings.

6.1 References

ASHRAE (1989). "Ventilation for Acceptable Indoor Air Quality," ASHRAE 62-1989, American Society for Heating, Refrigerating, and Air Conditioning Engineers, Atlanta, GA.

Colthorpe, K. (1990). "A Review of Building Airtightness and Ventilation Standards," Technical Note AIVC 30, International Energy Agency, Air Infiltration and Ventilation Centre, Coventry, Great Britain.

Mudarri, D.H. (1989). "Federal Directions in Indoor Air Quality," Proceedings of the Summer Annual Meeting of the American Institute of Chemical Engineers, Philadelphia, August 21–24, 1989, American Institute of Chemical Engineers, New York, NY.

National Institute of Building Sciences (1986). "Standards, Regulations and Other Technical Criteria Related to Indoor Air Quality," National Institute of Building Sciences, Washington, D.C.

Serraino, C. (1989). "Safety and Health Standards for Public Employees: Standards for Indoor Air Quality," Proposed New Rules: NJAC 12:100-14, New Jersey Department of Labor, Trenton, NJ.

U.S. Senate (1989). "Indoor Air Quality Act of 1989," Senate Bill S.657, Version of November 8, 1989, U.S. Senate, Washington, D.C.

WHO (1986). "Indoor Air Quality Research," Report on a WHO meeting, Stockholm, August 27–31, 1984, World Health Organization, Regional Office for Europe, Copenhagen.

CHAPTER 7

Future Needs in Indoor Air Quality Measurement Programs

Much progress has been made over the past few years in the development of successful approaches and techniques for the conduct of indoor air quality monitoring programs, but as in any developing field, there is always a need for improvements. The following sections present the authors' perceptions of where the field is going and ideas as to how each of the areas should be improved. This chapter is organized in a manner similar to that of the content in Chapters 2 through 5.

7.1 Planning an Indoor Air Quality Measurement Program

This area is in great need of improvement. As more emphasis is placed on total human exposure to air contaminants and developing quantitative relationships between exposure, dose and health risk, careful planning of the monitoring program and its statistical basis is of paramount importance. In planning such studies, the researchers must be thorough in reviewing the international literature to determine what is already known about the topics to be studied. For example, computerized databases are being developed by organizations such as the U.S. EPA on the results of indoor air quality and human exposure monitoring programs which should facilitate review.

While the number of studies to determine indoor concentrations of pollutants related to specific sources (e.g., NO_2 emissions from gas appliances) will probably decrease in the future, there will continue to be an increase in "sick building syndrome" (SBS) studies. There is a great need for the development of protocols and practical questionnaires for the cost-effective conduct of such studies. In addition, an improved working relationship between those conducting SBS studies and health professionals should be developed.

In the U.S., there is increasing interest among technical societies, governmental agencies, and consumer organizations to require certification of SBS investigators.

7.2 Ventilation Measurements

At present there are a variety of methods for measuring air exchange rates. Methods for measuring flows in ducts and exhaust flows from diffusers are readily available, but there is a need for flow measurement devices with greater sensitivity than the presently available hot wire anemometers used for measuring air movement in buildings. Smoke tubes can be used, but give only qualitative results on airflow patterns.

Tracer techniques are the most reliable methods for measuring air exchange rates, but current methods are usually complex and expensive to carry out. There is a need for increased simplification and automation of equipment and methods to make these methods more readily available. A new instrument introduced in 1989 by Brüel and Kjaer Instruments, Inc. capable of measuring a number of pollutants (primarily VOCs) at a number of locations simultaneously, also has the capability of introducing SF_6 into a space and simultaneously measuring its concentration.

The need for improved and less costly tracer methods is applicable to both large building studies and studies of individual structures where perfluorocarbon tracers are commonly used. In this connection, some effort should be devoted to the development of a mail-out tracer test package such that air exchange measurements can be made by home and building owners and occupants.

It has always been assumed that tracers such as SF_6 are completely conservative. This may be true chemically, but studies should be carried out to determine if this and other tracers adsorb on indoor surfaces, thus degrading their applicability to air exchange measurements.

Another serious research issue related to tracer and other types of ventilation studies is the evaluation of the "well mixed" assumption so often made. Studies are needed of the effect of the juxtaposition of air supplies and exhausts and barriers to airflow such as partitions on degree of mixing.

7.3 Measurement of Indoor Pollutants

Monitoring for indoor air quality has a relatively short history, but during this brief history development of new and improved methods has occurred rapidly and has progressed beyond the chart presented as Figure 4.1. Lewis (1989) has summarized recent developments in the laboratories of the U.S. EPA. One approach to defining research needs on methods development is to present what might be called the "ideal indoor air quality monitor" whose attributes include the following:

- High sensitivity.
- Quiet and unobtrusive.
- Ability to measure both short- and long-term average concentrations.
- Ability to measure a number of pollutants.

- Portable, suitable for personal monitoring.
- For direct reading instruments, ability to collect and store data for computer manipulation.
- Low power requirements.
- Inexpensive.
- Ease of operation by untrained workers.

Obviously, no monitoring device can possess all of these attributes, but the above list is of value in discussing progress made and future needs. The following discussion is organized in terms of the pollutants measured in indoor air quality monitoring programs and discussed in Chapter 4.

7.3.1 Criteria Pollutants

Among the criteria pollutants, the ones for which further developments are needed are CO, NO_2, and PM_{10}, but among these the current state of development of indoor and personal monitors for CO is already well advanced. SO_2, O_3 and lead are primarily outdoor pollutants. There eventually may be some increased emphasis on total personal exposures to these three outdoor pollutants in support of increased regulatory emphasis on total human exposure. Such reorientation may foster the development of personal monitors for these pollutants, but further developments along these lines are not a high priority item at this time.

Lewis (1989) reports that improvements have been made to the passive monitor for NO_2, reducing exposure times to hours instead of days. Evaluations should be made of these improvements by deploying these devices in indoor air quality studies representing a spectrum of NO_2 exposures, and comparing the results from established methods.

It is believed that short-term exposures to high concentrations of NO_2 indoors may represent a health hazard. Such short-term high concentrations have been measured indoors in homes with unvented gas stoves and heaters by many workers using continuous recording instruments designed for outdoor ambient monitoring. There is a need for smaller, portable instruments for use indoors and as personal monitors. Work on such systems is underway at the laboratories of the U.S. EPA (Lewis, 1989).

CO is most commonly measured in the outdoor ambient atmosphere and at fixed points indoors by continuous recording instruments based on non-dispersive infrared (NDIR) spectroscopic detection. Because such systems are too bulky and complicated for personal monitoring and most indoor air quality studies, U.S. EPA developed a small monitor called "COED" based on an electrochemical measurement principle (Ott *et al.*, 1986). Further experience with this system is needed with this device, and refinements to the data logging system will evolve as computer technology advances.

PM_{10} in the outdoor atmosphere is now beginning to be measured routinely throughout the U.S.A. As more emphasis is placed on total human exposure to pollutants, there will be a need for miniaturized PM_{10} monitors that can be easily carried by exposure subjects or worn by them. Already there are available personal monitors for respirable particulate matter (RPM). A research program funded by the U.S. EPA has produced an impactor-based personal sampler for collection of PM_{10} and RPM for subsequent analysis of elemental composition by X-ray fluorescence (Spengler *et al.*, 1989).

In addition to such personal monitors that provide integrated measurements, there is a need for rapid response, reliable, portable survey instruments for indoor monitoring that can provide accurate instantaneous data on a size selective basis. Currently available instruments based on light scattering are useful, but results are often hard to correlate with results from established methods using gravimetric analysis.

7.3.2 Radon and Decay Products

The U.S. EPA has issued guidance on the appropriate use of several different types of measurement devices for radon and radon progeny. These include continuous monitors which use scintillation cells, 3 to 5 day integrated charcoal canisters, alpha-track detectors, and grab sample scintillation cells. However, there is a need for field instrumentation which can measure aerosol concentrations and size distributions simultaneously with the attached and unattached radon progeny. A technique for passively monitoring radon progeny is also needed.

7.3.3 Formaldehyde

Formaldehyde as an indoor pollutant tends to be limited to specific situations such as mobile homes and other indoor spaces where large amounts of particle-board and paneling are used, and in those few remaining homes with UFFI. Nevertheless, it is still widely measured and often (usually mistakenly) implicated in SBS situations. Since formaldehyde emissions are principally from outgassing of construction materials (e.g., particle-board), their effect on indoor air quality is characterized by gradual increases (and decreases) over significant time-averaging periods rather than by short-term fluctuations. Thus, in spite of the implications of Figure 4.1, there does not appear to be a significant incentive for direct readout improved continuous analyzers for typical indoor surveys. Presently available techniques such as passive samplers and wet chemical methods appear adequate for present applications. Nevertheless, improved continuous samplers would be useful for monitoring formaldehyde emissions from building materials and furnishings.

7.3.4 Volatile Organic Compounds

At present there is more research being carried out in the development of improved methods for identifying and measuring VOCs than for any other class of pollutants. Presently there are adequate methods for measuring low level, long-term average concentrations of a wide range of VOCs in low concentrations by means of adsorption or whole air samplers with subsequent analysis with GC/MS with or without a cryogenic concentration step. However, only long-term averages can be measured, and the analytical procedures are involved and expensive. The greatest needs are for the following types of devices:

• Passive monitors that have the ability to distinguish between VOC species and that do not require expensive analytical procedures.
• Direct reading (e.g., color change) passive monitors for specific compounds or classes of compounds.
• Survey-type instruments such as compact, portable chromatographs that have the capability of providing on-site concentration measurements of important VOCs or groups of such compounds at low concentrations. Such devices would be particularly useful in SBS studies. The instrument mentioned in Section 7.2 developed by Brüel and Kjaer Instruments, Inc. and capable of measuring a number of VOCs at several indoor locations simultaneously should be evaluated as to sensitivity, specificity, and cost-effectiveness.

7.3.5 Environmental Tobacco Smoke

As environmental tobacco smoke is a mixture of numerous compounds, methods for monitoring different constituents have been developed. Currently, nicotine is the most common constituent of ETS to be measured. The development of the treated sodium bisulfate filter method appears to be an accurate and cost-effective sampling method. A passive nicotine sampler has been developed, but comparative tests are needed with other types of samplers.

Although nicotine is a unique tracer for ETS exposure, it is not the compound which causes the sensory irritation associated with environmental tobacco smoke. Aldehydes are considered to be one of the primary irritants in ETS. Portable, low cost methods for monitoring various species of aldehydes need to be developed.

7.3.6 Pesticides

The U.S. EPA has completed a large scale study of pesticides in 300 homes in two cities using a polyurethane foam (PUF) plug as a solid sorbent with a low volume sampling pump. The method appears to be acceptable for a broad range of pesticides; however, certain household pesticides, e.g., acephate and glysophate, have analytical problems. Further research is needed on these compounds. The sampling apparatus for the PUF plugs is not readily available beyond the research establishment. Greater commercial availability of the glass holder and PUF as a complete unit would enhance its usage. PUF has a breakthrough problem for semi-volatile compounds (SVOCs) with high vapor pressure. Further research is needed on multimedia sorbent traps for SVOCs.

7.3.7 Odor

Future work on odor as an indicator of indoor air quality must be directed at developing a better understanding of odor and other physiological sensations and their role in defining acceptable indoor air quality. Significant work has been done in the Scandinavian countries, but little attention has been given to this area by those developing indoor air quality policy in the U.S. Specific areas of needed research are:

- Development of more precise odor thresholds for indoor air pollutants and mixtures thereof.
- Use of odor panels or other groups of impartial judges that gather data in a wide range of buildings with and without indoor air quality problems. These techniques are being used in the Scandinavian countries, but they now should be pursued in the U.S. and other countries.
- Training of building owners, managers and occupants, and indoor air quality monitoring consultants in odor assessment.

7.3.8 Carbon Dioxide

Currently there is adequate instrumentation for continuous recording, fixed monitoring of CO_2 by NDIR. There is also portable equipment based on NDIR that can be used for SBS and other types of building surveys. However, there is a need for further miniaturization of this portable device and the addition of recorders or data loggers to facilitate fixed point and portable monitoring and the use of CO_2 decay data for calculating air exchange rates. Furthermore, a multiport CO_2 analyzer capable of measuring concentrations of CO_2 in several areas simultaneously (e.g., "complaint" and "non-complaint" areas) would be useful in SBS studies.

7.3.9 Bioaerosols

The research needs for bioaerosols are twofold: one relates to method development and the other to microenvironmental field studies. The two are so closely related that it is difficult to determine which should have first priority. Initially, the primary research should focus on developing a database on the types and levels of microorganisms found in office and residential settings. Current data are inadequate to compare results across studies, so this expanded research effort should use equivalent sampling and collection media. Such an expanded research effort would require an increase in the number of laboratories equipped to provide these analyses.

An expanded research program would lead logically to the second research need: the development of improved sampling methods. The improved methods should have the following attributes:

- Ability to be used in the field by personnel without highly specialized training.
- Collection media which will support the growth of a wide variety of bacteria and fungi.
- Extended sampling time. Such devices as the cascade impactor have sampling times on the order of 5 minutes. In field studies this very small sample of air is often considered representative of concentrations over extended periods, e.g., 24 hours and longer. A technique which could sample a very small volume of air at preset intervals would collect more representative samples.

The effects of dust mites on indoor air quality have been studied extensively in Europe, but little work on this area of research has been done in the U.S. Studies should be carried out in the U.S. that extend the European work and provide improved methods for assessing the importance of this source of indoor air contamination.

It is only recently that the significance of bioaerosols in the indoor environment became appreciated by those outside this rather narrow research field. Hopefully, this awareness will lead to more research in the field and a better understanding of the role bioaerosols play in such situations as the sick building syndrome.

7.4 Summary

The preceding sections identify a number of specific needs for improving indoor air quality monitoring techniques. All of these suggestions are based on the assumption that their use in indoor air quality monitoring studies will improve our knowledge of the importance of indoor exposure to air pollutants in defining human health. There is ample evidence that in past and present studies much of the data collected on indoor exposures is exploratory and anecdotal and provides little insight into the meaning of the data in the larger context of possible health effects. Indoor air quality researchers today and in the future have and will have an impressive array of highly sophisticated and sensitive monitoring techniques that are capable of measuring a contaminant of interest at miniscule concentrations in the indoor environment. Given these devices, there is the temptation to rush out and measure this compound in every imaginable microenvironment with little regard to interpreting the data in any meaningful way.

It is hoped that indoor air quality researchers in the future will spend time conceptualizing and designing rigorous indoor air quality studies that address specific hypotheses on exposures and suspected adverse health effects and less time on exploratory measurements because there is an instrument or method capable of measuring a specific pollutant.

7.5 References

Lewis, R.G. (1989). "Development and Evaluation of Instrumentation for Measurement of Indoor Air Quality," *Man and His Ecosystem, Proceedings of the 8th World Clean Air Congress 1989, The Hague,* Elsevier, Amsterdam.

Ott, W.R., Rodes, C.E., Drago, R.J., Williams, C., and Burman, F.J. (1986). Automated Data-Logging Personal Exposure Monitors for Carbon Monoxide," *J. Air Pollution Control Assoc.* **36**:883–887.

Spengler, J.D., Ozkaynak, H., Ludwig, J., Allen, G., Pellizzari, E.D., and Wiener, R. (1989). "Personal Exposures to Particulate Matter: Instrumentation and Methodologies P-TEAM," *Proceedings of the 1989 EPA/AWMA International Symposium on Measurement of Toxic and Related Air Pollutants,* Air and Waste Management Association, Pittsburgh, PA.

Appendices

These appendices present data collected in studies of pollutant concentrations in various microenvironments, including those indoors. This compilation was originally prepared by one of the authors under contract to the U.S. EPA, and it appeared in an EPA publication (U.S. EPA, 1988a). The version appearing here has been updated by the authors in 1990.

The appendices are organized in terms of the eleven pollutants measured. The references cited appear in the reference list at the end of Chapter 4. The following terms and abbreviations are used in the appendices except for Appendix G (Radon) which has its own terms and abbreviations:

$\overline{X}$ = arithmetic mean

$\overline{X}_g$ = geometric mean

Sd = standard deviation

Sd_g = geometric standard deviation

N = number of data points

AC = air conditioner

d = median aerodynamic diameter

BDL = below detectable limit

SE = standard error

r = correlation coefficient

r^2 = variance explained by linear regression

TWA = time weighted average

Appendix A Microenvironmental Field Study Data for Carbon Monoxide

Citation	Number and type of micro-environment	Location	Indoor source of interest	Sampling methods	Sampling period	Results (ppm) $\bar{X}$ Summer Day	Summer Night	Fall Day	Fall Night	Winter Day	Winter Night	Comments
Yocom et al. (1971)	Three pairs of buildings (public buildings, office buildings, private homes)	Greater Hartford, CT	Indoor/outdoor relationships	Intertech Infra-2 NDIR	Summer, fall, winter, 1969–70	PB 12.8 / PB 22.7 / OB 12.7 / OB 22.9 / PH 12.2 / PH 21.9	2.5 / 1.8 / 2.5 / 2.7 / 2.8 / 2.4	4.2 / 3.8 / 3.8 / 3.3 / 2.7 / 2.8	3.8 / 3.2 / 2.1 / 2.6 / 2.9 / 2.5	4.6 / 5.4 / 3.1 / 3.0 / 1.9 / 1.8	2.6 / 3.0 / 2.6 / 2.1 / 1.9 / 1.9	CO concentrations were higher with closer proximity to busy downtwon streets. CO readily penetrated all the structures; anomalies were readily relatable to source and ventilation variables. Submerged roadway or under-lying parking garage had little effect upon indoor air quality of associated structures.
						PB: Public building; OB: Office building; PH: Private home						
Petersen and Sabersky (1975)	Inside an automobile	Los Angeles, CA	None	CO monitor (Energetics Science)	Not reported	25 ppm CO not often exceeded. Highest concentration encountered: 45 ppm for a period of 3 minutes.						CO averages inside did not differ from those outside, despite short, high concentration peaks outside. No indication of inside CO sources.

Citation	Number and type of micro-environment	Location	Indoor source of interest	Sampling methods	Sampling period	Results (ppm) Indoor 1	2	3	4	$\bar{X}$ Outdoor	Comments
Wade et al. (1975)	Four houses	Hartford, CT	Gas stoves	Intertech NDIR	May 1973–Feb. 1974						Emissions from gas stoves contributed to indoor CO levels.
					House No.1						
					Spring 1973	—	3.9	3.6	3.6	3.0	
					Fall 1973 (1st half)	3.7	3.1	2.8	—	1.5	
					Fall 1973 (2nd half)	4.2	3.7	—	3.3	2.0	
					House No.3						
					Spring 1973	3.8	—	2.8	2.3	1.9	
					Fall 1973 (1st half)	6.8	5.6	4.4	—	3.0	
					Fall 1973 (2nd half)	6.2	5.8	—	4.8	2.2	
					House No.4						
					Winter 1974	7.9	7.9	7.2	—	2.1	
					1 = Kitchen, above stove 2 = Kitchen 3 = Living-room 4 = Bedroom						
Chaney (1978)	Inside car	Cross-country	Vehicle exhaust	Gas filter correlation monitor	March–April 1977	*Chicago/San Diego/LA Expressways* Traffic speed < 10 mph, CO > 15 ppm. Traffic speed = 0, CO = 45 ppm					Relatively small number of cars were responsible for high percentage of total CO emissions.
						Downtown New Orleans Traffic 3–5 mph CO = 2–50 ppm					High CO levels resulted from heavily loaded vehicles and vehicles ascending grade.
											Levels varied with traffic density and speed.

Study	Vehicles	Location	Source	Method		Car	X̄ Inside	X̄ Outside	Comments
Colwill and Hickman (1980)	Eleven new cars	London	Vehicle exhaust	Ecolyzers	Not reported	1	22.7	49.2	Internal levels depended on external changes, but changes greatly damped by ventilation system.
						2	30.5	55.8	Differences in internal CO levels more marked between vehicles than for different runs in the same vehicle (probably due to differences in the ventilation system).
						3	33.4	73.7	
						4	21.1	60.3	
						5	23.7	40.1	
						6	22.7	39.1	
						7	40.4	55.0	
						8	26.4	35.4	
						9	18.1	31.0	
						10	17.4	33.3	
						11	20.7	44.6	

Study	Vehicles	Location	Source	Method			MAX	N	Comments
Ziskind et al. (1981)	793 buses, 306 taxis, 127 police vehicles	Denver, CO; Boston, MA	Sustained-use motor vehicles	Glass stain detector tubes; Personal samplers; Continuous monitors; Tracer gas detectors	Not reported	Glass tube dosimeter			Potential is greatest for CO accumulation within a vehicle when doors, windows, and vents are all closed.
						Buses	37	11	CO intruded into passenger area of police cars and taxis through trunk, rear seat, and parcel shelf.
						Taxis	38	2	Principal CO sources were leaks from rear exhaust system, vertical inlet seal, engine exhaust manifold, and catalytic converters.
						Police cars	55	4	
						Personal sampler			
						Buses	50	6	
						Taxis	48	1	
						Police cars	34	4	
						Continuous monitoring			CO intruded most often into buses from rear emergency door seal, heater or windshield water hose, and fittings along the exhaust system.
						Buses	7	10	
						Taxis	17.4	2	

Citation	Number and type of micro-environment	Location	Indoor source of interest	Sampling methods	Sampling period	Results (ppm)	Comments
Brunekreef et al.(1982)	254 houses	The Netherlands	Gas-fired water heaters	Ecolyzer 2000 monitors	November–December 1980	Breathing height CO (see below)	Both vent and type of burner affected CO levels at breathing height.
Moschandreas and Zabransky (1982)	Residential sites and offices	Boston, MA	Spatial variation	Not reported	Not reported	Maximum CO concentration (see below)	Residences: differences were relatively small within unit. Statistical differences only in residences with gas appliances. Offices: statistical differences were attributable to the larger size of each building.
Ziskind et al. (1982)	Nine subjects commuting and at work	Los Angeles, CA	None	Personal monitors	Not reported	Activity (see below)	Subjects with greatest number of "high" CO days had longest commute. No significant elevations in the short term from turning on either heat or gas burners. Passive smoking at work significantly increased CO levels.

Brunekreef et al.(1982) — Breathing height CO

	N
<10	154
11–50	50
51–100	25
>100	17
Missing values	8

Moschandreas and Zabransky (1982) — Maximum CO concentration

Site	Kitchen	Bedroom	Living-room
1	9.96	8.73	7.36
2	5.28	5.28	5.28
3	10.64	10.01	12.54
4	6.06	4.54	6.06
5	4.70	4.70	3.82
6	11.32	2.58	2.70
7	7.19	4.44	1.94
8	11.81	12.95	12.57
9	10.29	10.12	10.62
10	8.73	9.55	10.04
11	7.36	0.97	0.97

Ziskind et al. (1982) — Activity

Activity	$\bar{X}$ Weekday	(N)	$\bar{X}$ Weekend	(N)
Home	4.6	(336)	4.0	(228)
Work	4.3	(551)	2.2	(6)
Commute	10.0	(418)	6.7	(47)
Errands	9.2	(62)	6.0	(50)
Leisure	6.5	(25)	4.2	(73)
Other	4.7	(78)	3.2	(11)

TWA with no smokers present 3.76 N = 438
TWA with smokers present 4.65 N = 135

Study	Setting	Location	Source	Method	Season	Details	Range	Comments
Traynor et al. (1982)	Environmental chamber and house	California	Kerosene-fired space heaters (white flame vs. blue flame)	Not reported	Not reported	Env. chamber white flame, blue flame; House white flame, blue flame	2–4, 2–14, <2, 2–7	White-flame convective heater emits less CO than blue-flame radiant heater. In small spaces and/or when air exchange rates are low, CO can exceed state and/or federal outdoor or occupational air quality standards.
Akland et al. (1984)	712 households—Washington; 900 households—Denver	Washington, D.C. and Denver, CO	Ambient pollution	Personal exposure monitors	Winter 1982–3	Eight-hour maximum results in Denver were approximately twice as high as levels found in Washington. Denver—10.7% of 8-hour maximum daily CO exposures >9 ppm. Washington—3.9% of 8-hour maximum daily CO exposure >9 ppm.		Only non-smokers in study. Personal CO exposures higher in microenvironments associated with motor-vehicles (i.e., commuting) and in high exposure occupations (i.e., truck drivers, construction workers).
Amendola and Hanes (1984)	13 service stations 2 dealerships	New England	Vehicle exhaust	Ecolyzer	Winter/Summer, year not reported	Warm weather, Cold weather	Range 2.2–21.6, 16.2–110.8	CO levels: Dealerships highest levels, small service stations intermediate levels, large service stations lowest levels. Cold weather CO levels significantly higher than warm weather due to closed ventilation.

184

Citation	Number and type of micro-environment	Location	Indoor source of interest	Sampling methods	Sampling period	Results (ppm) Avg.	Min.	Max.	$\bar{X}$	Sd	N	Comments
Coviaux et al. (1984)	Urban road tunnel	Paris center	Tunnel air pollution	Continuous monitor/ NDIR	May–June 1983							
	Technical room					6	2	15				CO levels strongly correlated with traffic intensity.
	Tunnel					30	11	41				Increased CO in technical room significantly correlated with smoking.
Flachsbart and Ott (1984)	558 commercial facilities	Five California cities and suburbs	General traffic patterns	Personal exposure monitors	November 1979–July 1980							
	Enclosed parking garages								27.7	12.5	10	CO levels relatively stable over time.
												CO lower on windier days.
	Settings attached to enclosed parking garages								6.1	2.9	7	Indoor CO levels were statistically but not substantially less than outdoors.
	All other indoor commercial settings								2.1	1.6	202	
	All outdoor settings								3.0	2.6	368	CO levels in parking garages were substantially greater than other types of indoor settings.
												CO levels inside settings varied as much within a given geographic location as they did between geographic locations.
												Average CO concentrations reported by fixed station monitors were higher than those reported by personal exposure monitors for both indoor and outdoor settings.

Flachsbart (1985)	Test vehicles	Honolulu. HI	Traffic patterns effect on intermodal CO concentrations	Personal exposure monitors/CO detector/integrator	November 1981–May 1982	Regular Auto (R) and Carpool (C) test vehicles	$\bar{X}$	CO exposure of commuters using priority lanes substantially lower than for those not using these lanes.
						Highway segment and travel mode		Higher speed on priority lanes contributed to reductions in CO exposure.
						Contraflow/with flow lane		
						R	11.8	
						C	9.7	
						Shared lanes		
						R	17.6	
						C	15.7	
						Regular Auto (R), Express Bus (EB), and High Occupancy Vehicles (HOV)		
						Highway segment and travel mode	$\bar{X}$	
						Undivided lanes		
						R	13.3	
						EB	5.6	
						R	12.0	
						HOV	11.8	
						Divided lanes		
						R	19.0	
						EB	7.8	
						R	15.3	
						HOV	8.5	
						Combined lanes		
						R	16.9	
						EB	6.6	
						R	14.4	
						HOV	10.4	

Citation	Number and type of micro-environment	Location	Indoor source of interest	Sampling methods	Sampling period	Results (ppm)	Comments
Lebret (1985)	Homes	Rotterdam/Ede, the Netherlands	Gas cookers, geisers	Ecolyzer 2000	Winter 1984	Real-time monitoring — Range: Kitchen 0–17.5; Living-room 0–8.7; Bedroom 0–3.5. Week-long monitoring — X̄: Kitchen 3.5; Living-room 1.7.	CO elevated in homes with gas cookers and unvented geisers. Kitchen levels higher due to peaks from use of gas appliances. Living-room values slightly higher in homes of smokers.
Sisovic and Fugas (1985)	Institutions housing sensitive population	Zagreb, Yugoslavia city center	None	Portable CO detector (GE) and Ecolyzer 2000 (Energetics Science Inc.)	Winter vs. summer, year not reported	See table below.	Indoor CO levels attributed to nearby traffic density, general daily pollution, seasonal differences, and day-to-day weather conditions.
Flachsbart and Ott (1986)	Nine high-rise buildings	California	None	Personal exposure monitors	January 1980–June 1984	See table below.	Underground parking garage was source in building with exceptionally high CO levels. Corrective actions proved successful.

Sisovic and Fugas (1985) Results (ppm):

Site	Winter X̄	Winter range	Summer X̄	Summer range
K1	3.4	1.7–4.6	1.8	0.6–3.4
K2	3.4	1.7–5.1	1.8	1.1–2.9
K3	6.0	3.4–13.7	4.9	2.9–6.9
K4	2.7	1.1–4.0	1.9	1.1–2.9
			1.7	1.1–2.9
K5	2.7	1.1–4.6	3.1	2.3–4.6
H1	4.2	1.1–7.4	2.4	1.1–4.6
H2	5.7	3.4–9.7	1.6	1.1–2.3
R1	3.4	1.7–5.7	2.4	1.7–4.6
R2	4.6	2.9–6.9	1.2	0.6–2.9
B	2.5	1.1–4.0		

Flachsbart and Ott (1986) Results (ppm): Four buildings exhibited elevated indoor CO concentrations; 1 building exhibited exceptionally high CO levels:

Floor of building	X̄ Before mitigation	X̄ After mitigation
15	5.5	0.7
11	9.1	1.0
6	9.3	1.2
3	9.0	1.5
1	8.1	0.8
Garage	36.2	7.9

					Winters	Wood heater type	$\bar{X}$	Peak	
Humphreys et al. (1986)	Research home	Tennessee	Wood stoves	Electronic analyzer	1983, 1984, 1985	**1985**			CO from non-airtight (NAT) heaters were generally higher than those from air-tight (AT) heaters.
						A	0.9	1.8	
						A	1.9	2.4	Catalytic AT heaters: Flue gas CO increased with increasing wood burn rate.
						A	0.9	1.4	
						A	1.6	2.8	
						B	0.9	1.3	
						1984			Conventional AT wood heaters: Flue gas CO decreased with increasing wood burn rate.
						C	3.0	9.6	
						D	3.6	29.6	
						C	2.0	5.6	
						1983			
						E	1.9	8.0	
						A	1.6	3.3	
						B	1.5	3.9	
						A	3.6	9.1	

Heater types
A = Catalytic, radiant heaters
B = Conventional, radiant heaters
C = Box-type, radiant heaters
D = Franklin-type, free-standing fireplace heater
E = Conventional, circular heater

Appendix B Microenvironmental Field Study Data for Nitrogen Dioxide

Part 1: Gas stoves

Citation	Number and type of micro-envronment	Location	Indoor source of interest	Sampling methods	Sampling period	Results ($\mu g/m^3$)				Comments
Wade et al. (1975)	Four homes	Hartford, CT	Gas stoves	Chemiluminescence, 2-week averages reported/season, three locations/home	Spring 1973 Fall 1973 Winter 1974	Kitchen: 60–140 Living-room: 28–75 Bedroom: 49–70 Outdoors: 32–101				Highest NO_2 levels in kitchen; peak concentrations related to stove use. Winter levels higher.
Spengler et al. (1979)	55 homes	Six cities: Portage, WI Topeka, KS Kingston, TN Watertown, MA Steubenville, OH St. Loius, MO	Not specified in advance	Bubbler, modified sodium arsenite	One every 6 days for May 1977– April 1978	City	$\bar{X}$ Gas	$\bar{X}$ Electric	$\bar{X}$ Outdoor	Indoor less than outdoor except in homes with gas stoves.
						Portage	20	8	11	
						Topeka	44	20	21	
						Kingston	—	13	19	
						Watertown	53	47	49	
						Steubenville	46	31	43	
						St. Louis	47	20	48	
						N=350–543				
Goldstein et al. (1979)	516 homes (houses and apartments, number not reported)	Middlesbrough, U.K.	Gas stoves	Diffusion tubes	Winter 1978	Gas kitchens: $\bar{X}$=212, SE=5.1 range: 9-598 N=428 Electric kitchens: $\bar{X}$=34, SE=4.5 range = 11-355 N=87				NO_2 levels generally higher than data from U.S. studies. Winter sampling, lower social economic status, stove type, humidity, and use of vents may have caused differences.

						$\bar{X}$ Gas	$\bar{X}$ Electric	
Dockery et al. (1981)	Nine homes and personal monitoring	Topeka, KS	Gas stoves	Passive dosimeters, Palmes tubes, 1-week exposure	June 1979, 4 weeks			Ambient $\bar{X}$ = 22.5 $\mu g/m^3$. Child had highest exposure in gas stove homes, but lowest in electric homes.
					Kitchen	59	16	
					Bedroom	37	14	
					Husband	36	20	
					Wife	41	17	
					Child	45	9	

						Median	Range	
Quackenboss et al. (1982)	Nine homes and personal monitoring	Portage, WI	Gas stoves	Diffusion— Palmes tubes	March 1981 3 weeks			Personal exposures for both types of homes not correlated with outdoor levels. Bedroom had highest correlation to personal concentrations for both types of homes.
					Gas			
					Kitchen	63	44–142	
					Outdoor	4	3–8	
					Personal	29	16–102	
					Bedroom	25	10–100	
					Electric			Bedroom values are the mean of two locations.
					Kitchen	8	2–17	
					Outdoor	12	4–17	
					Personal	12	8–29	
					Bedroom	4	0–10	

						N	$\bar{X}_g$	
Spengler et al. (1983)	137 homes (112 gas, 25 electric)	Portage, WI	Gas stoves both natural gas (NG) and liquid propane (LP)	Diffusion— Palmes tubes 1-week exposure	1 year July 1980 June 1981			N = number of 1-week measurements. Gas homes had higher concentrations in fall and winter. Thirteen homes exceeded NAAQS.
					Gas–NG			
					Kitchen	237	57	
					Bedroom	238	32	
					Outdoor	240	16	
					Gas–LP			No meaningful difference in NO_2 between NG and LP homes.
					Kitchen	568	56	
					Bedroom	568	30	
					Outdoor	555	11	
					Electric			
					Kitchen	174	7	
					Bedroom	172	5	
					Outdoor	173	12	

Citation	Number and type of microenvironment	Location	Indoor source of interest	Sampling methods	Sampling period	Results (μg/m³)	Comments
Clausing et al. (1984)	111 homes and personal monitoring	Watertown, MA	Gas stoves	Diffusion—Palmes tubes 1-week exposure	Winter 1982	**N / $\bar{X}$ / Sd** — Gas: Kitchen 42/80/34; Living-room 35/56/22; Outdoor 43/41/6; Personal 48/47/16. Electric: Kitchen 28/21/10; Living-room 28/17/5; Outdoor 48/47/6; Personal 71/24/7	Averages of two sampling periods. Regression models have $r^2 = 0.60$–0.83 for predicting personal exposure.
Hosein et al. (1986)	Homes and personal monitoring	Not reported	Gas stoves and air conditioning	Impinger, 24-hour average	Summer 9–29 days	**AC $\bar{X}_g$ / no AC $\bar{X}_g$** — Gas 175/182; Electric 71/82	AC appears to reduce concentrations but sample size small.
Goldstein et al. (1985)	Eighteen apartments (low income) and personal monitoring.	New York City	Gas stoves	Diffusion—Palmes tubes, 48-hour exposure	Winter 1984 30 days	**$\bar{X}$ / SE** — Kitchen 126/61; Living-room 82/35; Personal 45/20; Outdoor 75/23	NO_2 concentrations increased with height above floor. Personal less than outdoor is an unusual finding.
Lebret (1985)	Homes: Ede, 169 homes Rotterdam, 91 homes	Ede (suburban) and Rotterdam (urban), The Nether-	Gas stoves, unvented water heaters	Diffusion—Palmes tubes 1-week exposure	Winter 1981/82 1982/83	**N / $\bar{X}_g$ / Peak** — Ede: Kitchen 173/65/539; Living-room 173/36/166; Bedroom 172/28/151. Rotterdam: Kitchen 102/88/525; Living-room 102/47/212; Bedroom 102/41/173	N = number of weekly averages. No significant difference in concentrations between cities. Regression models using house characteristics explained 50–61% of the NO_2 variance.

Noy et al. (1986) — 35 homes and personal monitoring; Wageningen, The Netherlands; Unvented gas appliances; Diffusion—Palmes tubes, 1-week exposure; Winter 1984

Room	N	$\bar{X}$	Peak
Kitchen	35	106	331
Living-room	35	66	157
Bedroom	35	40	139
Ambient	—	41	—
Personal	52	77	157

Comments: NO_2 levels higher than U.S. studies. Peak values taken at time of appliance use.

Brunekreef et al. (1986) — Fifteen homes; Wageningen, the Netherlands; Unvented gas appliances; Diffusion—Palmes tubes, 1-week exposure; 1 year, 1982–1983

Room	$\bar{X}$ Winter	$\bar{X}$ Spring	$\bar{X}$ Summer	$\bar{X}$ Fall
Kitchen	74	77	56	61
Living-room	33	38	35	30
Bedroom	25	32	29	25

Comments: NO_2 concentrations relatively stable across seasons, except kitchen values lower in summer. All homes had gas stoves.

Goldstein et al. (1986) — 44 apartments; New York City; Gas stoves; Diffusion—Palmes tubes, 1-week exposure; Various seasons, 1982–4

Room	N	$\bar{X}$	Sd
Kitchen	799	111	40
Living-room	797	84	30
Bedroom	901	73	26
Outdoors	784	77	25

Comments: N = number of 1-week averages. Indoor concentrations are closer to European studies. 52% of kitchen values greater than NAAQS. Size of apartment and duration of stove use time important.

Part 2: Other Combustion Sources

Citation	Location	Number and type of microenvironment	Indoor source of interest	Sampling methods	Sampling period	Results ($\mu g/m^3$)	Comments
Ryan et al. (1983)	Vermont and Midwest	Various public buildings and 2 homes	Kerosene heaters	Diffusion—Palmes tubes, 4–7 days	Winter (year not reported)	Range School rooms 43–300 Church 77 Home 1 223–295 Home 2 26–64	Peak NO_2 concentrations in classrooms where heaters were used 10 hours/day. Home 1 used two heaters as primary source of heat; the other used one heater as a supplement.
Kim et al. (1986)	Seoul, South Korea	20 offices	Kerosene heaters	Diffusion—Palmes tubes, 4–5 day exposure	Winter 1984	$\bar{X}$ / Sd Kerosene 162 / 89 Electric 38 / 9	Smokers increased NO_2, but increase not statistically significant.
Yarmac et al. (1987)	Atlanta, GA	22 homes	Unvented gas space heaters	Chemiluminescence	February–April 1985, 4 days/site	Median / Range 2 heaters N=1: 231 / 0–1377 1 heater N=15: 94 / 0–3340 No heater N=8: 38 / 0–736 Ambient: 57 / 0–207	All except one "no heater" home had gas stove. NAAQS exceeded at 12 homes.
Good et al. (1982)	Richmond, VA	90 homes	Cigarette smoke	Diffusion—Palmes tubes, 7 day exposure	August 1980 and February 1981	$\bar{X}$ / Sd Smoker 21.3 / 11.5 Non-smoker 17.5 / 8.4	No significant difference in summer. Incremental exposure is estimated at 3–4 $\mu g/m^3$.

Appendix C Microenvironmental Field Study Data for Sulfur Dioxide

Citation	Number and type of micro-environment	Location	Indoor source of interest	Sampling methods	Sampling period	Results ($\mu g/m^3$)	Comments
Biersteker et al. (1965)	Sixty homes	Rotterdam, The Netherlands	None—outdoor SO_2	Drechsel bottle with H_2O_2	Winter 1964 7 days/site	Range: 0–246	Average indoor SO_2 concentration 20% of average outdoor concentration. Multiple linear regression analysis of age of house, fuel, smoking habits, and outdoor SO_2 gave $r^2=0.16$.
Wilson (1968)	Laboratory	Imperial College, London	Decay of SO_2 indoors	Wosthoff U3S—electrical conductivity	Winter 1968	Range: 49–154 (estimated)	Half life of SO_2 indoors depends on surface material for removal. When walls were coated with Na_2CO_3, half life decreased from 40–60 min. to 7 min.
Yocom et al. (1970)	Two public buildings, 2 office buildings, 2 homes	Hartford, CT	None outdoor SO_2	Conductimetric	1969–70; Summer Fall Winter	$\bar{X}$ Public bldg: —, 55, 162; Office bldg: —, 31, 99	$\bar{X}$ Home: 55, 44, 21. Indoor less than outdoor SO_2 concentrations except in home with leaky coal furnace. Indoor values ranged from 6–95% of outdoor values.
Andersen (1972)	One school	Arhus, Denmark	None—Outdoor SO_2	Thorin—Spectrophotometric	$7\frac{1}{2}$ months	Range: 10–60 $N=11$, 24-hour val;ues	Indoor concentrations averaged 51% of outdoor concentrations.

Citation	Number and type of micro-environment	Location	Indoor source of interest	Sampling methods	Sampling period	Results ($\mu g/m^3$) City	$\bar{X}$	Peak conc.	Comments
Spengler et al. (1979)	55 homes	Six cities: Portage,WI Topeka,KS Kingston,TN Watertown,MA St. Louis,MO Steubenville, OH	None— Outdoor SO_2	Bubbler— spectrophoto- metric	1 year, every 6 days, $N=330$–543 1977–1978	Portage Topeka Kingston Watertown St. Louis Steubenville	5 1 2 5 15 1	10 5 2 10 27 27	SO_2 reduced indoors from 10–90%. Highest reduction in air conditioned homes; lowest in WI and KS where outdoor levels were low.
Stock et al. (1985)	Twelve homes	Houston, TX	Outdoor SO_2	Pulsed fluorescence	May–October, 1981; 1 week/site $N=2455$ hours	Indoor Outdoor	13.3 7.3	13.9 13.1 (Sd)	Outdoor SO_2 less than indoor, an apparent anomaly.

Appendix D Microenvironmental Field Study Data for Ozone

Citation	Number and type of micro-environment	Location	Indoor source of interest	Sampling methods	Sampling period		Results (ppm)	Comments
Sabersky et al. (1973)	Two office buildings, one home	Pasadena, CA	None; outdoor O_3	Chemiluminescence (Dasibi)	Summer 1971 1 day	Office building 1 0–0.21 Office building 2 0–0.18 Home 0–0.21		Indoor less than outdoor O_3 concentrations. Peak outdoor concentration = 0.25. Indoor pattern of variability followed outdoor trend. House indoor = 0.7 × outdoor and peaks lagged by 1 hour.
Hales et al. (1974)	Laboratory building	Pasadena, CA	None; outdoor O_3	Chemiluminescence (Dasibi)	October 1973 1 day	Indoors: 0–0.15		Indoor less than outdoor O_3 concentration peaks lagging and reduced in magnitude.
Sutton et al. (1978)	Five homes	Minneapolis, MN Denver, CO Troy Hill, NY	Electrostatic air cleaner	Chemiluminescence	Winter 1972 1–2 days/site	Indoors: Air cleaner: 0.002–0.008 No air cleaner: 0.002–0.003		Indoor less than outdoor O_3 concentrations. Air cleaner does not contribute significant O_3. Furnace fans remove outdoor O_3.
NAS (1986b)	Airline cabin	Not reported	Upper atmospheric O_3	Not reported	Spring $N = 5600$ "observations"	Inside cabin: 0.05–0.625 Outside cabin: 0.05–1.0		Inside cabin concentrations correlated to outside concentrations at $r = 0.82$. FAA O_3 standard (1985) peak = 0.25, 3-hour average = 0.10.
Stock et al. (1985)	Twelve homes	Houston, TX	None; outdoor O_3	Chemiluminescence	May–October 1981 1 week/site $N = 2332$ h	Indoor Outdoor	$\bar{X}$ 0.003 0.022 / Sd 0.006 0.025	Outdoors, NAAQS exceeded for 126 hours. Indoor values do not show outdoor peaks. Air conditioning usage not reported.
Contant et al. (1986)	Thirty subjects for personal monitoring	Houston, TX	None; outdoor O_3	Chemiluminescence	8 hours/day for 2 days/ participant	Indoor Outdoor Vehicles (from personal monitoring)	Mean 0.011 0.052 0.023 / Median 0.005 0.042 0.014	Data used to refine O_3 exposure model. Houses with no air conditioning had higher O_3 concentrations.

Appendix E Microenvironmental Field Study Data for Respirable Particles

Citation	Number and type of micro-environment	Location	Indoor source of interest	Sampling methods	Sampling period	Results ($\mu g/m^3$)	Comments
Dockery and Spengler (1977)	22 homes	Watertown, MA	None, but influence of smokers noted	Cyclone separation/filtration, 24 hours, $d = 3.5 \ \mu m$	Summer 1975 16 days, Winter 1976 12 days	$\bar{X}_g \simeq 21$, $\bar{X}_g \simeq 21$	Indoor concentrations greater than outdoor concentrations.
Moschandreas et al. (1980)	Three homes	Boston, MA	Wood stoves	Cyclone separation/filtration, 24 hours, $d = 3.5 \ \mu m$	Winter, 2 weeks (year not reported)	$\bar{X}$ during wood burning Site A 49 B 160 C 68	BaP, CO, and TSP reported. Air infiltration measured. All pollutants higher during wood-burning periods. Site B had a smoker.
Ju and Spengler (1981)	Four homes	Boston, MA suburbs	None; 2 homes had light smokers	Cyclone separation/filtration, 24 hours, $d = 3.5 \ \mu m$	December 1979 25–30 days/site	$\bar{X}$ for all residents = 15–27, Range = 8–72	Indoor greater than outdoor levels. Monitoring in different rooms within each residence indicated generally good mixing of particles indoors.
Spengler et al. (1981)	55 homes	Six cities: Portage,WI Topeka,KS Kingston,TN Watertown,MA St. Louis,MO Steubenville, OH	None, but influence of smokers noted	Cyclone separation/filtration, 24 hours, $d = 3.5 \ \mu m$	1 year, once every 6 days 1977–8	City / $\bar{X}$ / Peak conc. Portage 20 43 Topeka 21 42 Kingston 48 50 Watertown 30 70 St. Louis 43 140 Steubenville 43 58	Indoor greater than outdoor levels at all cities except Steubenville, OH, the most industrialized city. SO_4, SO_2, and NO_2 reported.

Study	Building	Location	Source	Method	Time period	Values	Comments
Yocom (1982)	Elementary school	Ohio	None, but effect of different ventilation rates studied	Dichotomous sampler, 24 hours $d = 2.5$ μm	Not reported	Range: 11–19	Indoor less than outdoor levels. Highest indoor level associated with highest ventilation rate.
Sexton et al. (1984)	24 homes	Waterbury, VT	Wood stoves	Cyclone separation/ filtration, 24 hours $d = 3.5$ μm	Winter 1981–2 24 samples per site	$\bar{X} = 25$, $Sd = 3$; $\bar{X}_g = 22$, $Sd_g = 1.7$; Range = 6–69 ($N = 163$ samples)	Indoor greater than outdoor levels. Sampling in different rooms showed particle levels well mixed, except when kerosene heaters present. Kerosene heater homes were 10 μg/m³ higher than homes with wood stoves. Ventilation rates showed no strong relationship to indoor particle levels.
Sega et al. (1984)	Five office buildings	Urban area, Yugoslavia	None, but smokers noted	Cyclone separation/ filtration, particle size not reported	Winter, Summer 2-week period, year not reported	$\bar{X} = 98$ ($N = 60$) $\bar{X} = 40$ ($N = 25$)	Winter indoor values consistently greater than summer indoor values.
De Bortoli et al. (1985)	Five apartments, 9 detached houses	Northern Italy	None, but smokers noted	TSI, piezobalance (instantaneous values) particle size not reported	Various times over 1 year, 1983.	Range = 40–140 peak value at downtown site $\bar{X} = 77$ N not reported	Primary focus organics. Effect of smokers not pronounced. Data reported for individual sites. Ventilation rates reported.
Stock et al. (1985)	Twelve homes	Houston, TX	None	Dichotomous sampler, 12 hours $d = 2.5$ μm	May–October 1981 1 week/site ($N = 156$)	$\bar{X} = 21.3$ $Sd = 16.4$	Indoor greater than outdoor levels. Other pollutants reported: O_3, NO_2, SO_2, CO, pollen, and spores.

Citation	Number and type of micro-environment	Location	Indoor source of interest	Sampling methods	Sampling period	Results ($\mu g/m^3$)	Comments
Hosein et al. (1986)	Homes, number not reported	Not reported	Various combustion sources, air conditioning, and carpets	Cyclone separation/ filtration, particle size not reported	Summer, year not reported	No AC, No smokers ($N=11$) $\bar{X}_g = 32$ $Sd_g = 1.84$ > 1 smoker ($N=25$) $\bar{X}_g = 70$ $Sd_g = 1.64$	Particle levels higher in homes with carpet and air conditioning. Impact of smokers evident.
TVA (1985)	One research house	Chatta-nooga, TN	Wood stoves	Dichotomous sampler 12 hours $d = 2.5$ μm	Winter 1984	Overall range = 12–310 Baseline (stove not operating $\bar{X}$ = 28; stove operating $\bar{X}$ = 94	TSP also reported. Percentage TSP respirable 50%, lower than expected.
Humphreys et al. (1986)	One research house	Chatta-nooga, TN	Wood stoves	Dichotomous sampler 24 hours and 12 hours $d = 2.5$ μm	Winter 1984	Airtight, $\bar{X}$ = 27–38, Sd = 2–11 Non-airtight, $\bar{X}$ = 44–91, Sd = 11–82.	CO, NO_2, and PAH also reported. Non-airtight stoves generate higher levels of CO and particles than airtight stoves.

Appendix F Microenvironmental Field Study Data for Lead

Citation	Number and type of microenvironment	Location	Indoor source of interest	Sampling methods	Sampling period	Results ($\mu g/m^3$) $\overline{X}$	Comments
Yocom et al. (1970)	Two public buildings, 2 office buildings, 2 homes	Hartford, CT	None; outdoor aerosol	Modified Hi-Vol and atomic absorption	Summer, Fall, Winter (N = 9–21 days, each 24 hours) 1969–70	**Public bldg** 0.86 1.84 1.17 / **Office bldg** 0.28 0.51 0.99 / **Home** 1.41 1.28 0.74	Indoor less than outdoor concentrations. Maximum indoor concentration was 2.04 $\mu g/m^3$. Air conditioned air-rights structure did not have elevated concentrations.
General Electric (1972)	Two apartment buildings	New York City	None; Not reported aerosol		Year not reported / Heating season	**Air rights** 1.42 (2nd) 1.58 (roof) / **Street canyon** 1.45 (11th) 1.53 (18th)	All means exceed NAAQS. Lower floors greater than higher floors. Indoor greater than outdoor only at higher floors.
					Non-heating season	1.51 (2nd) 1.53 (roof) / 1.91 (11th) 2.19 (18th)	Only street canyon building showed seasonal differences.
Azar et al. (1975) (cited in U.S. EPA, 1986)	Office building, taxi cabs, industrial plants	California Pennsylvania Florida Wisconsin	Auto exhaust	Not reported	Not reported	$\overline{X}_g$ / Sd_g PA-taxi cabs 2.59 / 1.16 FL-plant 0.59 / 2.04 WI-plant 0.61 / 2.39 CA-taxi cabs 6.02 / 1.18 CA-offices 2.97 / 1.29	Concentrations are from personal exposure monitors. Higher taxi cab exposures in L.A. may be related to ambient concentrations being higher than in Philadelphia.
Halpern (1978)	Apartment building (<10 yrs old), apartment building (<40 yrs old), 2 museum locations (<75 yrs old)	New York City	None; outdoor aerosol	Serial filtration for respirable and non-respirable fraction and atomic absorption	Summer Year not reported N = 18–22, 2 hours each	Apt bldg (newer): $\overline{X}$ = 0.21, Sd = 0.13 range = 0.12–0.40 Museum (main floor): $\overline{X}$ = 0.38, Sd = 0.19 range = 0.79 Museum (storeroom): $\overline{X}$ = 0.63, Sd = 0.18 range = 0.45–0.98 Apt bldg (older): $\overline{X}$ = 0.36, Sd = 0.15 range = 0.14–0.51	Indoor means less than outdoor means across all sites. Total number of days sampled not reported. Non-air-conditioned sites had greater concentrations than air-conditioned sites.

Citation	Number and type of micro-environment	Location	Indoor source of interest	Sampling methods	Sampling period	Results ($\mu g/m^3$)	Comments
Moschandreas et al. (1978)	Eighteen residences (apartments and houses)	Various cities	None; outdoor aerosol	Filtration for total particulate matter and atomic absorption	18 months, 16 days per site, 24 hours each 1976-77	$\bar{X} = 0.36$, $Sd = 0.24$ range = 0.1-1.1 (N = 63, 4-day average concentrations)	Indoor means less than outdoor for all but nine 4-day periods, not consistent across sites.
Moschandreas et al. (1979)	Two conventional homes, one test house	Baltimore, MD and Denver, CO	None; outdoor aerosol	Filtration ("streaker") and PIXE	Summer 1976 7 days at 2-hour intervals	$\bar{X}$ for 7 days = 0.30 (estimated from a plot for Denver residence)	Data from only one house presented. Outdoor mean value = 1.17 $\mu g/m^3$. Data indicates possible re-entrainment of outdoor aerosols indoors.
LBL (1980)	Two test houses	Ames, IA Carroll County, MD	None; outdoor aerosol	Dichotomous sampler, XRF	One house winter Ames—range ($N \simeq 15$ days) One house spring Carroll—range	$\simeq 0.01$-0.04 $\simeq 0.01$-0.04	Indoor less than outdoor for all samples.
Diemel et al. (1981)	101 homes	Arnhem, The Netherlands	Lead smelter and soil contamination indoors	Filtration, d < 3-4 μm and atomic absorption	Season not reported, 24 hours 1978	$\bar{X}_g = 0.26$ $\bar{X} = 0.27$ range = 0.13-0.74	Also sampled Pb in dustfall and floor dust. Indoor less than outdoor Pb concentrations, but indoor TSP greater than outdoor TSP. Correlation between indoor Pb and fine floor dust = 0.49. No association between indoor and outdoor Pb levels.
Tosteson et al. (1982)	Eleven homes	Topeka, KS	None; outdoor aerosol	Cyclone separation, filtration and atomic absorption	Spring 1979 N=59, 24 hours each 1979	$\bar{X} = 0.09$ Median = 0.08 Std. error = 0.009	Also reported Al and Fe. Outdoor and personal samples collected. For mean Pb concentrations personal greater than outdoor greater than indoor.

Appendix G Microenvironmental Field Study Data for Radon-22 and Decay Product Concentrations Measured in U.S. Residences. (Adapted from A.V. Nero et al., "Distribution of Airborne ^{222}Radon Concentrations in U.S. Homes", *Science*, vol. 234 (1936) pp. 992–993.)

Location	No. of Houses	^{222}Rn concentrations[†]			Measurement protocol[*]		
		AM	GM (pCi/liter)	GSD	Tech.	Per.	Seas.
Washington	138	1.11	0.71	2.48	TD	3m	W
Oregon	90	1.27	0.88	2.27	TD	3m	W
Montana	20	1.87	1.32	2.34	TD	3m	W
Idaho	17	3.56	1.30	2.94	TD	3m	W
San Francisco, CA	29	0.45	0.35	2.19	GS	I	S
Las Vegas, NV	3	1.06	0.79	2.47	TD	ly	Yr
Colorado Springs, CO	16	2.54	1.87	2.34	TD	4–5m	WSp
Fargo, ND	11	7.62	5.69	2.39	TD	4–5m	WSp
Wisconsin	50	1.45	1.10	2.17	TD	3–4m	W
Houston, TX[‡]	103	0.59	0.47	1.94	TD	3–6m	SF
Portland, ME	11	0.70	0.44	2.98	TD	4–5m	WSp
New York	27	1.57	0.89	2.65	TD	2–9m	SF
Rochester, NY	8	0.93	0.50	4.08	CR	2w	W
Northeastern NY	9	1.33	0.78	3.12	TD	~6m	FWSp
NY City Area	11	0.49	0.45	1.56	PM	1w/seas	2Yr

Location	No. of Houses	^{222}Rn concentrations†			Measurement protocol*		
		AM	GM (pCi/liter)	GSD	Tech.	Per.	Seas.
New Jersey	9	1.55	1.38	1.73	PM	1w/seas	2Yr
Princeton, NJ	11	2.47	1.09	3.95	TD, GS	~5m	WSp
Pittsburgh, PA‡	122	2.37	1.45	2.35	TD	1y	Yr
Philadelphia, PA area	31	2.64	1.84	2.32	AC	3d/seas	WS
Damascus, MD	41	3.91	3.04	2.11	AC	3d/seas	WSp
East Tennessee‡	40	3.04	2.0	2.5	TD	2 × 3m	WF
Charleston, SC	20	0.58	0.47	2.12	TD	4–5m	WSp
Butte, MT‡	179	6.39§	3.46§	3.03	RP	1w/seas	Yr
Grand Junction, CO‡	62	1.57§	1.39§	1.61§	RP	1w/1m	Yr
Santa Fe, NM	6	1.23§	0.94§	2.25§	RP	1–16d	Yr
Farmington, NM	6	0.83§	0.74§	1.66§	RP	1–12d	Sps
New Mexico	9	3.29	2.31	2.45	GS	I	Sp
Chicago, IL‡	44	1.02	0.5	3.3	GS	I	F
Central Maine	70	2.97	1.66	2.56	TD	6–8m	1–2Yr
Lewiston, NY	10	0.52	0.50	1.31	PM	1–2w/seas	2Yr

					Tech	Per	Seas
Middlesex, NJ	15	0.54	0.52	1.39	PM	1–2w/seas	2Yr
Canonsburg, PA	8	1.30	0.94	2.15	PM	1–2w/seas	2Yr
Eastern Pennsylvania	36	7.57	3.29	3.41	TD	2–8m	WS
Maryland	58	3.80	1.90	3.69	GS	I	SpSF
Oak Ridge, TN	14	1.69§	1.28§	2.24§	GS	I	W
Raleigh, NC	10	0.80	0.52	2.87	GS	I	F
Florida (non-mineralized)	29	0.81§	0.73§	1.60§	RP	1w/seas	Yr
Florida (mineralized)	4	2.75§	2.14§	2.51§	RP	1w/seas	Yr

* Measurement protocol is indicated by symbols. Technique (Tech): TD = Track-Etch™ detector, GS = grab sample, CR = continuous radon monitor, PM = PERM (passive environmental radon monitor), AC = activated carbon integrating device, RP = RPISU (radon progeny integrating sampling unit). Period (Per): (#)y,m,w, or d = continuous number of years, months, weeks or days; I = instantaneous; /seas means measurement performed each season. /lm means each month. Season (Seas): W = winter, Sp = spring, S = summer, F = fall, Yr = spans four seasons.

† Where individual data are available, preference is given to main floor living-room or to averaged living area values, from which the arithmetic mean (AM), geometric mean (GM), and geometric standard deviation (GSD) are calculated; otherwise these parameters are taken directly from the original paper or (in a few cases) derived from crude histograms.

‡ No data on individual houses; only statistical data were obtained.

§ 222-Rn decay-product concentration was measured.

Appendix H Microenvironmental Field Study Data for Formaldehyde: Mobile Home Studies

Citation	Number and type of micro-environment	Location	Indoor source of interest	Sampling methods	Sampling period	Results (ppm)	Comments
Garry et al. (1980)	275 mobile homes	Not reported	Building materials	Impinger/ chromotropic acid	February– June 1979	$\bar{X}$ February 0.65 March 0.38 April 0.23 May 0.58 June 1.0	HCHO levels changed with season (indicating temperature/humidity effects). Newer mobile homes tend to have higher HCHO levels.
Hanrahan et al. (1984)	65 mobile homes	Wisconsin	Building materials acid	Impinger/ chromotropic	Not reported	Indoor Range = <0.10–0.80 Median = 0.16 $N = 65$ Outdoor $\bar{X} = 0.04$ $Sd = 0.03$ $N = 33$	Positive dose–response relation-ship between prevalence of eye discomfort and formaldehyde concentrations.
Sterling et al. (1984, 1986)	Two mobile homes	Texas	Temporal factors, propane vs. electric as cooking source, materials testing	Impinger/ chromotropic acid	October 1982 November 1983	Month / $\bar{X}$ No temperature control / $\bar{X}$ Temperature control Oct 0.53 0.46 Nov 0.67 0.41 Dec 0.94 0.70 Jan 0.43 0.34 Feb — 0.47 Mar 0.33 0.21 April 0.38 0.38 May 0.68 0.44 June 1.00 0.35 July 0.55 0.33 Aug 1.52 0.32 Sept 1.08 0.33 Oct — — Nov 0.49 0.29	HCHO decreased with time, HCHO increased with indoor temperature at rate of ∼ 50% per 8°C. Slight decrease with time for temperature controlled home.

Reference	Homes	Location	Source	Method	Dates	Results	Findings
Hanrahan et al. (1985)	137 mobile homes	Wisconsin	Building materials	Impinger/chromotropic acid	Not reported	Range = <0.01–2.84 Distribution mean = 0.46 Median = 0.39 $\bar{X}_g = 0.37$ $N = 138$ Monthly $\bar{X}$: Homes <3 yrs old = 0.54 Home >3 yrs old = 0.198	Only 10% of homes studied showed significant variation in measurement due to placement of monitors; concentrations 80% higher in kitchen. Higher HCHO levels in homes <3 yrs old.
Lamm (1986)	Mobile homes	Washington Minnesota Wisconsin		Impinger/chromotropic acid	Varied	Washington $\bar{X} = 0.6, N = 187$ Minnesota $\bar{X} = 0.4, N = 109$ Wisconsin (A) $\bar{X} = 0.66, Sd = 0.65, N = 65$ Wisconsin (B) $\bar{X} = 0.24, Sd = 0.23, N = 65$	HCHO levels inversely proportional to age of home.
Marchant et al. (1986)	One mobile home	Saskatoon, Saskatchewan	Building materials	Impinger/chromotropic acid	Not reported	Daytime living-room = 0.67 Night-time bedroom = 0.68	Levels uniformly distributed indicating evenly distributed source. Major emission source was particle-board panelling used throughout.
Norsted et al. (1985)	443 mobile homes	Texas	Age of home	Colorimetric detector tubes	April 1979–May 1982	Range = <0.5–8.0 27% of homes ≤ 1 year old had $\bar{X} \geq 2$ ppm 11.5% of homes >1 year old had HCHO concentrations ≥ 2 ppm.	Inverse relationship found between home age and HCHO levels.
Sexton et al. (1985)	663 mobile homes	California	Building materials	Passive dosimeters	July–August 1984	$\bar{X}_g$, kitchen = 0.071 $Sd = 2.0, N = 611$ $\bar{X}_g$, bedroom = 0.071, $Sd = 2.1$ $\bar{X}_g$, overall = 0.072, $N = 663$ Range = 0.010–0.464 1957–80 mobile homes $\bar{X}_g = 0.061$ vs. 1981–84 mobile homes $\bar{X}_g = 0.080$	Variation in HCHO levels gradually decreased over time. Approximately one third of homes had HCHO concentrations higher than 0.1 ppm (ASHRAE standard).

Citation	Number and type of micro-environment	Location	Indoor source of interest	Sampling methods	Sampling period	Results (ppm)	Comments
Lin et al. (1979)	Energy-efficient research houses	Various regions in the U.S.	Effect of ventilation rates	Impinger/chromotropic acid, pararosaniline and MBTH	1978-79	Energy efficient research houses. With ventilation rates of 0.3 ACH or less, indoor HCHO levels can exceed 0.1 ppm. Outdoor concentrations were typically 0.016 or less.	Indoor greater than outdoor HCHO levels. Homes and office trailers have indoor HCHO levels that can exceed known health effect levels.
	Public school	Columbus, Ohio				Public school and large medical center. Similar indoor/outdoor concentrations both being well below 0.1 standard (probably due to high ventilation rates).	
	Medical center	Not reported					
	New office trailers	Long Beach, CA				New office trailers — Range: Indoor 0.1-0.2; Outdoor <0.008	
Dally et al. (1981)	100 buildings	Wisconsin	HCHO levels related to age of structure and presence of smokers investigated	Impinger/chromotropic acid	January 1978-November 1979	N / X̄ / Range: All homes 100 / 0.35 / <0.10-3.68; Mobile homes 65 / 0.47 / <0.10-3.68; UFFI homes 14 / 0.10 / 0.10-1.09; UFFI wood product conventional homes 13 / 0.10 / 0.10-0.92; Travel trailers 2 / 1.06 / 0.48-1.63; Office buildings 2 / 0.44 / 0.38-0.49; UFFI foamed and UFFI wood product conventional homes 2 / 0.22 / 0.17-0.28	HCHO concentrations decreased with increasing age of structure's building materials. Smoking did not significantly increase indoor HCHO concentrations.
Schutte et al. (1981)	16 homes, 11 homes	Boulder, CO; Wisconsin	Relationship between UFFI stability and weather conditions	Impinger/chromotropic acid	January-March 1980	Colorado range = <0.010-0.021; N = 10 UFFI homes; X̄ = 0.039; Outside = non-detectable-0.009; Range = 0.0-0.025; N = 3 control homes; X̄ = 0.019. Wisconsin range = 0.028-0.144; N = 11; X̄ = 0.079	No observable correlation between age of UFFI and HCHO concentrations. Higher indoor HCHO concentrations in Wisconsin vs. Colorado

Reference	Sample	Location	Source	Method	Dates	Results	Comments
Hawthorne et al. (1984)	40 homes	Eastern Tennessee (Oak Ridge, West Knoxville)	Building materials	Passive membrane sampler/pararosaniline	April–December 1982	Houses > 5 yrs $\bar{X}$ = 0.04, N = 22 Houses <5 yrs $\bar{X}$ = 0.08, N = 18 $\bar{X}$ Spring 0.062 Summer 0.083 Fall 0.040	25% of homes exceeded HCHO levels of 0.10. HCHO concentrations increase with temperature. Inverse relationship between age of house and HCHO levels. Elevated levels found in new homes not containing UFFI.
Dement et al. (1984)	Modular offices and new office building	Not reported	Building materials	Impinger/chromotropic acid pararosaniline	1981–83	Range / Peak / N Modular Unoccupied w/o vent — 0.33–0.78, 0.80, 24 Occupied before improvement — 0.06–0.23, 0.29, 39 Occupied after improvement — <0.05–0.08, 0.08, 6 New Initial (9/81) — 0.15, 0.34, 18 Interim (1/82–3/82) — 0.10, 0.40, 155 Most recent (8/82–6/83) — 0.08, 0.11, 10	Wall paneling most significant source. Ammonia fumigation appears to be effective means of reducing emissions-finding most significant HCHO emissions based on short-term monitoring; long-term study needed.
Broder et al. (1986)	450 UFFI homes vs. 225 control homes	Toronto	UFFI	Impinger/chromotropic acid	1983, 2 measure-/ments/site, 12 months apart	Control $\bar{X}$ Indoor = 0.035 Outdoor = 0.005 UFFI (no action) Indoor = 0.045 Outdoor = 0.005 UFFI (removal) Indoor = 0.044 Outdoor = 0.005	More abnormal health indicators in all UFFI homes but greater in UFFI removal homes.

Citation	Number and type of micro-environment	Location	Indoor source of interest	Sampling methods	Sampling period	Results (ppm)				Comments
						N	$\bar{X}$	Sd	Range	
Stock and Mendez (1985)	78 homes	Houston, Texas	Relationship of indoor/outdoor levels, type of structure, and age of building materials	Impinger/chromotropic acid	Summer 1980					Indoor concentrations elevated due to indoor sources. Type of structure and age of building materials affect indoor HCHO concentrations, as well as air exchange rates.
						Conventional house 36	0.04	0.03	<0.008–0.14	
						Energy eff. house 7	0.07	0.02	0.04–0.11	
						Apartment 19	0.08	0.07	0.02–0.27	
						Condominium 10	0.09	0.06	<0.008–0.29	
						Energy eff. condominium 3	0.18	0.03	0.15–0.20	
Sullivan et al. (1986)	26 UFFI homes; 6 control homes	London, Ontario	UFFI seasonal variations	Impinger/chromotropic acid; Pro-Tek C-60 passive dosimeters	August, October, December 1982; January, April, July 1983					Evidence of seasonal effects for the UFFI-responder classification but not the other three.

(a)

	$\bar{X}$ Control CTA/Pro-Tek	$\bar{X}$ UFFI responder CTA/Pro-Tek	
Aug	0.06	0.08	0.07
Oct	0.06	0.06	0.06
Dec	0.05	0.05	0.06
Jan	0.04	0.04	0.04
Apr	0.04	0.05	0.04
Jul	0.04	0.10	0.07

(b)

	$\bar{X}$ UFFI non-responder CTA/Pro-Tek	$\bar{X}$ UFFI removed CTA/Pro-Tek	
Aug	0.04	0.03	0.03
Oct	0.04	0.04	0.04
Dec	0.05	0.04	0.03
Jan	0.04	0.02	0.03
Apr	0.04	0.03	0.03
Jul	0.05	0.03	0.03

(a) UFFI homes showing obvious seasonal change in HCHO concentrations
(b) UFFI homes not showing obvious seasonal change in HCHO concentrations Complaint Mobile Homes

Syrotynski (1986)	2318 residential settings	New York State	Building materials	Impinger/ chromotropic acid	Sept. 1979– Jan. 1984	Complaint mobile homes $\overline{X} = 0.18$ ($N = 161$) Peak = 1.61	Complaint mobile homes have higher HCHO concentrations.
						Permanent residential UFFI homes w/complaints $\overline{X} = 0.06$ ($N = 1954$) Peak = 0.49	Complaints come from both UFFI and non-UFFI residences.
						Permanent residential complaint homes without UFFI $\overline{X} = 0.08$ ($N = 1953$) Peak = 2.60	
						Permanent residential non-complaint homes without UFFI $\overline{X} = 0.03$ ($N = 50$) Max. = 0.11	
Syversen et al. (1985)	Seven energy efficient homes versus three reference houses	Trondheim, Norway	Building materials, house furnishings	ORBO-22 absorption tubes (Supelco)	Not reported	Energy efficient homes <0.0001; all results <0.0001 Reference homes 0.0002–0.0006	Chipboard used as structural element in house is an important HCHO source.
							Other sources (e.g., textiles) less important.

Appendix I Microenvironmental Field Study Data for Volatile Organic Compounds

Part 1: Study designs

Citation	Number and type of Microenvironment	Location	Indoor source of interest	Sampling methods	Sampling period	Comments
Johansson (1978)	Two school-rooms and outdoors	Sweden	Human occupancy	Poropak Q tubes with thermal desorption and GC/MS or GC-FID analysis	Not reported	Certain compounds associated with human metabolism. More VOCs and higher concentrations indoors than outdoors. Concentrations of nine compounds estimated from a plot.
Mølhave and Moller (1979)	39 older homes, seven new unoccupied homes	Denmark	Building materials	Charcoal tubes, desorption with dimethylformamide and GC-FID or GC/MS analysis, 25-hour samples	Spring and Summer 1976 N = 38	Total VOC concentrations in new homes greater than in older homes. VOC classes identified were: 34% alkylbenzenes, 25% alkanes, 23% terpenes, 18% other. Forty compounds identified, but concentrations not reported.
Seifert and Abraham (1982)	Fifteen apartments (indoors, outdoors); traffic intersection	West Germany	Household products, auto exhaust	Charcoal pads—passive sampler with GC analysis, collection period > 1 day.	Not reported N = 15 indoors N = 5 outdoors N = 48 traffic	Four VOCs identified. Concentrations in kitchen and living-room are similar. Toluene associated with newsprint and magazines.
Wallace et al. (1985)	150 homes, indoors and outdoors	New Jersey (and other states)	Household products	Tenax GC™ tubes with thermal desorption and GC/MS analysis, two 12-hour samples collected/site	Various times from 1981–4	Eleven VOCs found in most samples. Levels indoors greater than outdoors.
Wallace (1986)	Personal exposure and outdoors	Los Angeles, CA (and other cities)	Personal activities	Tenax GC™ tubes with thermal desorption and GC/MS analysis, two 12-hour samples/individual	February and May 1984 N = 117 N = 71	Nineteen VOCs identified by personal monitors.

Reference	Description	Location	Sources studied	Method	Time	Findings
De Bortoli et al. (1986)	Fourteen homes, one office building	Italy	Building materials, household products, combustion products	Tenax GC™ tubes with thermal desorption and GC-FID or GC/MS analysis, charcoal tubes for halogenated compounds with GC-ECD analysis. Sampling time = 4–7 days/site with low flow pump.	Winter 1983–4, $N = 2$ samples/site	35 compounds identified and concentrations reported. Levels indoor greater than outdoor. Paints and wood impregnants identified as sources.
Lebret et al. (1986)	319 homes stratified by age	The Netherlands: Ede—134 post-war 96 <6 yrs; Rotterdam—89 pre-war	Building materials and household products	Charcoal tubes, 5–7 day integrated samples collected, GC-FID and GC/MS analysis for confirmation of a limited number of samples	Winter 1981–2	45 VOCs identified. Concentrations indoor greater than outdoor.
Hawthorne et al. (1986)	Forty homes	Tennessee	Building materials, household products, and combustion products	Tenax GC™ tubes with GC and GC/MS analysis. Samples collected for 6 hours and 24 hours	April–December 1982	Seventeen VOCs quantified. Levels indoor greater than outdoor. Levels reported by season and for homes with and without combustion sources.
Wallace (personal communication) (1987)	Three "new" buildings one "old" building three indoor and one outdoor sample at each location	Washington, D.C.	Building materials and furnishings	Tenax GC™ tubes with thermal desorption and GC and GC/MS analysis	Various times, 1983–5	Indoor concentrations of some VOCs in new buildings up to 100X outdoor values. Half lives of various VOCs estimated at 2–4.8 weeks. Indoor–outdoor differences in old buildings less than in new buildings, except for three compounds.

Appendix I (*continued*)

Part 2: Results for compounds reported in more than one study

Concentrations ($\mu g/m^3$)

Compound	Johansson (1978) X̄ In*	Johansson (1978) X̄ Out	Mälhave and Moller (1979) X̄ In	Seifert and Abraham (1982) X̄ In	Seifert and Abraham (1982) X̄ Out	Seifert and Abraham (1982) Traffic	Wallace et al. (1985)‡ X̄g In	Wallace et al. (1985)‡ X̄g Out	Wallace (1986)§ X̄ Personal	Wallace (1986)§ X̄ Out	De Bortoli et al. (1986) X̄ In	De Bortoli et al. (1986) X̄ Out	Lebret et al. (1986)‖ Median In	Lebret et al. (1986)‖ Median Out	Hawthorne et al. (1986)# X̄ In
Aromatic															
Benzene	4	5		15	29	†	12	4.1	18	16	52	20	5	3	11
Ethylbenzene			130	15	13		6.4	2.5	11	9.7	27	7.4	2	0.4	
m-p-Xylene	12**	8		29	28	100	16	8.3	28	24	89	24	10	3	44
o-Xylene				9			5.3	2.8	13	11	26	8.7			
Toluene	20	10	90	62	35	147					128	40	43	5	62
Styrene							1.5	0.55	3.6	3.8	36	11	5	0.7	
Trimethylbenzene															
n-Propylbenzene															
Butylbenzene													<0.3		
m,p-Dichlorobenzene						5.1	1.0	18	2.2	55	<5	1	<0.6	<0.3	
Naphthalene											11	<0.6	<0.3		18
1-Methylnaphthalene												2	<0.3	3	21
Limonene											140	1	45	10	
α-Pinene			480						4.1	0.8	102	<1.6			
Aliphatic															
n-Hexane											71	14	3	2	6
n-Heptane											16	5.1	2	1	9
n-Octane											14	2.4	1	<0.3	12
n-Nonane	2	1							5.8	3.9	27	2.1	6	<0.3	14
n-Decane	2	1	710						5.8	3.0	92	3.1	14	0.4	17
Undecane									5.2	2.2	80	<2	9	0.4	8
Dodecane									2.5	0.7	20	<1.7	4	<0.3	2
Tri-decane											3.1	<1.3	2	<0.3	4
n-Tetradecane					2		<0.3	8							
n-Pentadecane					2		<0.3	2							
n-Hexadecane					1		<0.3	4							

Chlorinated								
Chloroform	3.3	0.55	1.9	0.7	1.9	<1		
1,1,1,-Trichloroethane	19	3.4	96	34	21	11		
Trichloroethylene	2.6	1.4	7.8	0.8	18	7.5	<2	<2
Carbon tetrachloride	1.8	0.8	1.0	0.6	6.3	7.0	<4	<4
Tetrachloroethylene	6.3	2.1	16	10	18	14	<2	<2
Oxygenated								
Acetone	20	5				3.9		
Ethanol	65	6	220			6.1		

* Values are from occupied classrooms.

† Range 12–193, no mean reported.

‡ Indoor night-time samples, N.J. only $N = 86$.

§ Personal samples, L.A. California. $N=115$.

|| New homes <6 yrs old (Ede).

Summer season, all homes.

** Total xylenes.

Appendix I—Part 2 (continued)

Compound	Concentrations (μg/m³)*		
	Sheldon et al (1988)		
	New office 2 t = 1–6 wks	New office 2 t = 2–4 months	Maximum outdoor concentrations
Aromatic			
Benzene	3	5	5
Ethylbenzene	51	5	2
m-p-xylene	59†	19†	62†
o-xylene			
Toluene			
Styrene	3	3	4
Trimethylbenzene	110	13	3
n-Propylbenzene	9	2	BDL
Butylbenzene			
m,p-Dichlorobenzene	BDL‡	3	BDL
Naphthalene			
Limonene			
α-pinene	14	25	BDL
Aliphatic			
n-Hexane			
n-Heptane			
n-Octane			
n-Nonane			
n-Decane	440	15	4
Undecane	210	34	2
Dodecane	150	24	1
Tri-decane			
n-Tetradecane			
n-Pentadecane			
n-Hexadecane			

Chlorinated			
Chloroform	13	39	10
1,1,1,-Trichloroethane	BDL	2	1
Trichloroethylene			
Carbon tetrachloride	BDL	2	1
Tetrachloroethylene			
Oxygenated			
Acetone			
Ethanol			

* See original citation for data on other buildings.

† Total xylenes.

‡ Less than detection limit.

216

Appendix I (*continued*)

Part 3: Results for compounds reported in only one study

Concentrations ($\mu g/m^3$)

Compound	Wallace (1986) X̄		De Bortoli et al. (1986) X̄		Lebret et al. (1986)* Median		Hawthorne et al. (1986) X̄
	In	Out	In	Out	In	Out	In
3-Methylpentane					2	1	
2-Methylhexane					2	1	
3-Methylhexane					1	0.9	
Methylhexane					1	0.6	
Cyclohexane						0.4	
Dimethylcyclopentane					0.5	0.3	
2-Butanone			7.9	3.8			
Trichlorofluoromethane			39	10			
1,2-Dichloroethane	0.5	0.2					
Methyl isopropylbenzene					1	<0.3	
Butylbenzene					1	<0.3	
o-m-p-Methylethylbenzene					2	<0.3	
Chlorobenzene					<0.4	<0.4	
o-Dichlorobenzene	0.4	0.2					
Trichlorobenzene					<0.8	<0.8	
p-Dioxane	0.5	0.456					
Cumene							2
Mesitylene							7

* New homes <6 years old (Eds).

Appendix J Microenvironmental Field Study Data for Nicotine Concentrations

Citation	Number and type of microenvironment	Occupancy	Ventilation	Sample time	Levels ($\mu g/m^3$) Mean	Range
Harmsen and Effenberger (1957) in U.S. Surgeon General, 1986	Train	Not given	Natural, closed	30–45 min		0.7–3.1
Hinds and First (1975*) in	Train	Not given	Not given	$2\frac{1}{2}$ h	4.9	
U.S. Surgeon General, 1986	Bus	Not given	Not given	$2\frac{1}{2}$ h	6.3	
	Bus waiting room	Not given	Not given	$2\frac{1}{2}$ h	1.0	
	Airline waiting-room	Not given	Not given	$2\frac{1}{2}$ h	3.1	
	Restaurant	Not given	Not given	$2\frac{1}{2}$ h	5.2	
	Cocktail lounge	Not given	Not given	$2\frac{1}{2}$ h	10.3	
	Student lounge	Not given	Not given	$2\frac{1}{2}$ h	2.8	
Badre et al. (1978)	Six cafés	Varied	Not given	50 min		25–52
	Room	18 smokers	Not given	50 min	500	
	Hospital lobby	12–20 smokers	Not given	50 min	37	
	Two train compartments	2–3 smokers	Not given	50 min		36–50
	car	3 smokers	Natural, open	50 min	65	
			Natural, closed	50 min	1010	
Weber and Fischer (1980)†	44 offices	Varied	Varied	140 × 3 h	0.9±1.9	13.8 (peak)

Citation	Number and type of microenvironment	Occupancy	Ventilation	Sample time	Levels ($\mu g/m^3$)	
					Mean	Range
Hammond et al. (1987)	Chamber	Four smokers[†]	2.5 ACH	4 h	99	77–111
	Office	Non-smokers[†‡]	Not given	8 h	14.6	3.1–25.7
		Smokers[‡]	Not given	8 h	36.5	25.1–48
	Railroad workers	Non-smokers	Outdoors	8 h	0.1	0.0–0.2
		Smokers	Outdoors	8 h	5.8	0.9–16.1
Thompson et al. (1989)	36 restaurants	Varied	Not given	1 h	5.4 [§] (3.5)[§]	0.5–37.2

* Background levels have been subtracted
† Control values (unoccupied rooms) have been subtracted
‡ Personal Sampling
§ Geometric Mean

Appendix K Microenvironmental Field Study Data for Pesticides

Citation	Number and type of microenvironment	Location	Indoor source of interest	Sampling methods	Sampling period	Results ($\mu g/m^3$)	Comments
Wright and Leidy (1980)	One commercial pest control building, 2 service vehicles, 6 food preparation areas	North Carolina	Airborne pesticide residues	Not reported in this publication	September 1977	**Pest control bldg** — Office / Storage Chlorpyrifos 0.126 / 0.220 Diazinon 0.163 / 0.284 Malathion BDL / 0.128 DDVP 0.041 / 0.617 **Food preparation areas** Chlorpyrifos only @ $t = 0$: 0.020–1.49 @ $t = 24$ hr: 0.004–0.361 **Vehicles** $\bar{X}$ Chlorpyrifos 0.096 Diazinon 0.130 Malathion 0.077 DDVP 0.110	Some pesticides transported from storage areas by ventilation system. Concentrations in food preparation decayed rapidly. Air movement may increase concentrations. Levels in vehicles vary by where equipment is carried.
Jackson and Lewis (1981)	Test-room in laboratory	North Carolina	Eighteen pest control strips	Porapak R and glass fibers with GC-ECD or GC-FPD, 2-hour samples	Various times up to 30 days after application of strips	Peak / Day 30 Propoxur 0.80 / 0.70 Diazinon 1.40 / 1.21 Chlorpyrifos 0.25 / 0.16	Maximum concentrations reached between days 7-15. Background levels <0.03.
Livingston and Jones (1981)	~498 apartments	Randolph Air Force Base, TX	Chlordane applied to subslab	Chromosorb 102 and anhydrous Na_2SO_4, GC-ECD, analysis 2-hour samples,	January 1979, 1980	1979 survey: 0.4–293; 60% apartments >1.0 1980 survey: BDL–38; $\bar{X} = 1.9$; 77% apartments had detectable concentrations	Decrease in 1980 reflects effect of cleaning units with high concentrations. No association of indoor concentrations to year of application. Some migration to 2nd floor noted.

220

Citation	Number and type of microenvironment	Location	Indoor source of interest	Sampling methods	Sampling period	Results ($\mu g/m^3$)	Comments
Wright et al. (1981)	49 dormitory rooms	North Carolina State University, Raleigh, NC	Airborne application of insecticides	Impingers with hexylene glycol or PUF with GC analysis	Daily for 3 days following applications	**Peak / Day 4 / Decay** Acephate 1.3 / 0.3 / 77% Bendiocarb 7.7 / BDL / — Carbaryl 1.3 / 0.01 / 99% Chlorpyrifos 1.1 / 0.30 / 73% Diazinon 1.6 / 0.40 / 75% Fenitrothion 3.3 / 0.50 / 85% Propoxur 15.4 / 0.70 / 95%	Most peak concentrations occurred immediately after application. Relative magnitude of concentrations do not follow vapor pressures. Airborne concentrations not related to percent active ingredient.
Gebefugi and Korte (1984)	Three houses	Munich, FRG	Wood preservatives	Trapping media not reported, GC and GC/MS analysis	Season not reported, 1974	**House A / B** Lindane 0.18 / BDL Pentachlorophenol 0.6 / 0.5	Measured concentrations in various home materials and blood. Carpets and floor dust had elevated concentrations.
Jurinski (1984)	Seven buildings	Not reported	Pesticides—subterranean application	Chromosorb 102 or iso-octane in impinger with GC-ECD analysis, 2-hour samples	Not reported	Air concentrations for House C not reported **Bldg / Heptachlor / Chlordane** A 1.6–2.1 / 0.1–0.2 B 0.3–4.3 / <0.1–0.5 C 0.1–3.1 / <0.1–0.6 D 1.6–14.8 / 0.4–0.8 E 1.4 / <0.1–0.8 H 0.75–1.3 / <0.75 I 0.9–6.9 / <0.1–3.2 N = 1–9	Higher heptachlor concentrations due to higher vapor pressure. Basement areas can have higher concentrations than other areas.
Lewis et al. (1986)	Nine residences	Not reported	Household usage of pesticides	Polyurethane foam, diethyl ether extraction, and GC/ECD or GC/MS analysis	August 1985 1 24-hour sample/site	**$\bar X$ indoor / $\bar X$ outdoor / $\bar X$ personal** Chlorpyrifos 2.4 / 0.059 / 1.9 Diazinon 1.4 / 0.11 / 0.85 Chlordane 0.51 / 0.058 / 0.68 Propoxur 0.042 / 0.0034 / 0.10 Heptachlor 0.088 / 0.016 / 0.060	22 pesticides were identified, but those reported here were the most common.

Index

Bioaerosols (*cont.*)
  analysis, 132–133
  characteristics, 124–128
  database, research needs, 175
  filtration devices, 130, 131–132
  health effects, 129
  indoor concentrations, interpretation,
      133–134
  measurement, 128, 130–133
    bioassays, 132, 133
    biochemical assays, 132
    cascade impactors, 131
    centrifugal impactor, 131
    gravity samplers, 130
    high-volume electrostatic sampler, 131
    immunoassays, 132–133
    inertial impactors, 130, 131
    liquid impingers, 131
    sieve impactors, 131
  method development, research needs,
      175
  microenvironmental field studies,
      research needs, 175
  samplers, 130
  sampling, 28
    research needs, 176
  size range, 125
  sources, 124–128
Bivariate analysis, 32
Breakthrough volumes, for volatile organic
    compounds, 100
Bubblers, pesticide sampling, 112
Building materials, emissions
    measurement, 151–152, 154
Buildings
  infiltration, 36
  mechanical ventilation, 18–19
  permeability, 7
  pollutant sources, 6
  sampling sites, 19
  ventilation, 36
Butanol dilution, odor threshold response
    curves, 119

Carbon dioxide
  characteristics, 122–123
  in homes, 124
  indoor concentrations, 2, 122
  indoor measurement, 123–124
  measurement, 123, 147, 150
  multiport analyzer, research needs, 175

  sampling, 27
  sources, 122–123
  as tracer gas, 47, 123
  WHO standard, 164
Carbon monoxide
  in buildings, 69
  catalytic oxidation monitor, 67
  characteristics, 66
  COED monitor, 172
  electrochemical monitor, 67
  from unvented indoor combustion
      appliances, 69
  gas filter correlation method, 67
  and heart disease, 21
  indoor air research, 172
  indoor concentrations, 63, 68–69
  inside motor vehicles, 69
  measurement, 67–68, 147–150
  microenvironmental field studies,
      179–187
  in motor vehicle exhaust, 69
  non-dispersive infrared (NDIR) analyzer,
      67
  passive monitors, 67
  personal monitor, 20
  sampling, 27
  sources, 66
  as tracer gas, 47
  WHO standard, 164
Carpets, formaldehyde, 94
Cascade impactor, 131
Catalytic oxidation monitor, carbon
    monoxide, 67
Centrifugal impactor, 131
Chamber studies, 12–13
Charcoal canisters, radon, 88
Cigarettes, nitrogen dioxide, 72–73
Clean Air Act, 2
COED monitor, for carbon monoxide, 172
Combustion gases, measurement, 147–150
Construction materials, 19
Continuous analyzers, for criteria
    pollutants, 27–28
Continuous monitors, 9–10
Continuous radon monitors, 88
Continuous scintillation cell, radon, 91
Continuous working level monitor, radon
    progeny, 88, 91
Criteria pollutants, 2
  continuous analyzers, 27
  research needs, 172–173

Index compiled by Judith A. Douville